Cambridge Elements ☰

Elements in Decision Theory and Philosophy
edited by
Martin Peterson
Texas A&M University

EVOLUTIONARY GAME THEORY

J. McKenzie Alexander
*The London School of Economics and Political
Science*

CAMBRIDGE
UNIVERSITY PRESS

Shaftesbury Road, Cambridge CB2 8EA, United Kingdom

One Liberty Plaza, 20th Floor, New York, NY 10006, USA

477 Williamstown Road, Port Melbourne, VIC 3207, Australia

314–321, 3rd Floor, Plot 3, Splendor Forum, Jasola District Centre,
New Delhi – 110025, India

103 Penang Road, #05–06/07, Visioncrest Commercial, Singapore 238467

Cambridge University Press is part of Cambridge University Press & Assessment,
a department of the University of Cambridge.

We share the University's mission to contribute to society through the pursuit of
education, learning and research at the highest international levels of excellence.

www.cambridge.org
Information on this title: www.cambridge.org/9781108713474
DOI: 10.1017/9781108582063

First published 2023

A catalogue record for this publication is available from the British Library.

ISBN 978-1-108-71347-4 Paperback
ISSN 2517-4827 (online)
ISSN 2517-4819 (print)

Evolutionary Game Theory

Elements in Decision Theory and Philosophy

DOI: 10.1017/9781108582063
First published online: March 2023

J. McKenzie Alexander
The London School of Economics and Political Science

Author for correspondence: J. McKenzie Alexander, j.alexander@lse.ac.uk

Abstract: Evolutionary game theory originated in population biology from the realisation that frequency-dependent fitness introduced a strategic element into evolution. Since its development, evolutionary game theory has been adopted by many social scientists and philosophers to analyse interdependent decision problems faced by boundedly rational individuals. Its study has led to theoretical innovations of great interest for the biological and social sciences. For example, theorists have developed a number of dynamical models which can be used to study how populations of interacting individuals change their behaviours over time. This introductory Element covers the two main approaches to evolutionary game theory: the static analysis of evolutionary stability concepts; and the study of dynamical models, their convergence behaviour, and rest points. This Element also explores the many fascinating and complex connections between the two approaches.

Keywords: evolutionary game theory, bounded rationality, evolutionary dynamics, cooperation, local interactions

ISBNs: 9781108713474 (PB), 9781108582063 (OC)
ISSNs: 2517-4827 (online), 2517-4819 (print)

Contents

1 Introduction
1.1 Rock, Paper, Scissors (Lizard, Spock)

In 1996, Barry Sinervo and Curtis Lively, two scientists from the Department of Biology and the Center for the Integrative Study of Animal Behavior at Indiana University, published a paper describing surprising population behaviour concerning the species *Uta stansburiana*, a.k.a. the common side-blotched lizard. In this species of lizard, males occur in three different behavioural types, identified by coloured blotches on their necks. The first type, with an orange throat, aggressively defends large territories containing multiple female lizards. The second type, with a dark blue throat, differs in that it is both less aggressive and prone to defending smaller territories containing fewer female lizards. The third type of male, with a yellow throat, is visually similar to females and – importantly – does not defend any territory at all.

What Sinervo and Lively found was that, over time, these three types of males exhibited an interesting pattern of variation in how frequent each type was in the population. Initially, the orange-throated males increased in number: aggressively defending large swathes of territory enabled them to mate with more female lizards, resulting in their having more offspring than other types. Eventually, though, a tipping point was reached. Defending larger territories meant that a single orange-throated male had to divide his time policing a wider area, and was not always able to prevent the yellow-throated males (which resembled females, remember), from invading their space and mating with the female lizards. This led to the yellow-throated type increasing in number. However, after some time yet another tipping point was reached. The yellow-throated type became vulnerable to the blue-throated type. Why? The fact that the blue-throated type would aggressively defend a small territory mean that they were able to prevent the yellow-throated type from sneaking in. This gave the blue-throats a fitness advantage, causing their type to increase in number. However, once the blue-throated types were populous enough, their less aggressive nature made them vulnerable to the orange-throated type, who would expand their territory into areas previously occupied by the blue-throats. And, then, the cycle would begin again.

Now consider one of the first games young children learn to play: Rock-Paper-Scissors. In this game, each child counts "one–two–three" in unison and then makes the shape of either rock, paper, or scissors with one hand. The rules determining the winner are well known: paper covers rock (so paper wins), but rock breaks scissors (so rock wins), and scissors cut paper (so scissors win). The point to note is that no single choice is the best regardless of what the other person chooses: each choice can either win or lose, depending on what the other person picks. Rock-paper-scissors is thus a game of *strategy*, even if not a very interesting one.[1]

Children have been playing rock-paper-scissors for over two thousand years. The first written description of the game dates from around 1600, when the Chinese author Xie Zhoazhi, writing during the period of the Ming dynasty, stated that the game's origins went as far back as the Han dynasty, which spanned from 206 BC to 220 AD. Back then, instead of rock-paper-scissors the objects of choice were frog-snake-slug, but the game was otherwise the same.

When children play rock-paper-scissors, they understand the rules of the game. Each child knows that whether or not they will win depends on both their choice and the other person's choice. In contrast, the common side-blotched lizard does not understand the structure of their reproductive environment. The lizards do not "choose" their throat colour or their behaviour in any way remotely similar to how children choose rock, paper, or scissors. But the *strategic* aspect to what is going on in both cases is essentially the same. What the example of the lizards demonstrates is the central topic with which this Element is concerned: how the interaction between individual behaviours in an evolutionary setting is such that natural selection *poses*, and then *solves*, problems of *strategy* even though none of the creatures involved are *rational*. This is what gave rise to the field known as *evolutionary game theory*.

The name 'evolutionary game theory' is composed of two parts. The root of the expression, 'game theory' refers to the formal study of problems of strategy, a interdisciplinary study spanning mathematics, economics, computer science, and other disciplines. 'Evolutionary' is an adjective which serves to qualify the particular questions and methods one is interested in when studying those problems of strategy. We'll see in a moment how, exactly, the idea of evolution, an idea fundamental to population biology, became intertwined with the study of strategic problems. But for now the following observation may help: in both cases, the idea of "the best thing to do" makes no sense in the absence of further context. The best way for prey to avoid a predator depends on how the predator pursues the prey. The best way to play chess depends on the skill level of your opponents. And what is even more interesting is that sometimes the "rules" of the game can themselves change, like when a predator invents a new method of pursuit, or when the offside rule is introduced in football. Fans of the TV show *The Big Bang Theory* will be familiar with the extended version of

---

[1] That said, the fact that the World Rock Paper Scissors Association proudly bills itself as a 'professional organization' for Rock-Paper-Scissors players around the world suggests that some people take the game *very* seriously. See https://wrpsa.com.

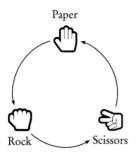

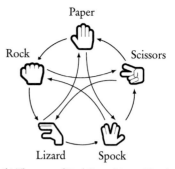

(a) The game of Rock-Paper-Scissors.

(b) The game of Rock-Paper-Scissors-Lizard-Spock.

Figure 1 A comparison of the game of Rock-Paper-Scissors (which is found in the animal kingdom) and the extended game of Rock-Paper-Scissors-Lizard-Spock (which is not). An arrow pointing from strategy S_1 to strategy S_2 means that S_1 wins when played against S_2. (Icons courtesy of Font Awesome.)

Rock-Paper-Scissors, originally invented by Sam Kass and Karen Bryla, which adds two additional moves – Lizard and Spock – to the children's game (see Figure 1b). No known species has competitive behaviour which matches the description of this new game, but let's wait and see what evolution produces in the future.

1.2 Game Theory

In 1945, the economist Oskar Morgenstern and the polymath John von Neumann published their seminal book *Theory of Games and Economic Behavior*. The title might strike some as curious: how can a element be both about *games* and *economics*? Surely economics, what Thomas Carlyle called the 'dismal science', is about as far removed from the study of games as possible?

The connection between games and economics derives from an important point about the kinds of choices involved in both. To see this, consider the difference between the kinds of choices a farmer makes when deciding to plant crops, and the kinds of choices a chess player makes when deciding which piece to move. In both cases, there is an optimisation problem. The farmer needs to determine the optimal time to sow the fields, taking into account expectations about future weather. The chess player needs to determine the optimal move to make, given the particular configuration of pieces on the board. But an important difference exists between the two types of optimisation problems. Although we might say that the farmer is trying to 'outwit Nature', that is really just a manner of speaking: Nature does not actually respond to the farmer's actions. Nature does not anticipate that the farmer is going to sow his or her crops at a certain time, and then choose not to rain out of spite (even though it might often feel like that). The weather unfolds in the same way it would have, regardless of what the farmer chooses to do. This is what is known as a problem of *parametric* choice: the farmer is deciding what to do given various parameters, some of which are known for certain, and some of which are either uncertain or unknown altogether. In the case of the chess player, her optimal choice is complicated by the fact that she is interacting with another person. Her opponent responds not only to previous moves as indicated by the position of chess pieces on the board, but to *beliefs* about what is likely to happen in the future. If she sees her opponent make an apparently imprudent move, the thought process that will trigger is readily imagined: 'Was that a mistake? Or is this an attempt to trick me into making a move whose future consequences I've not yet fully considered?' This is what is known as a problem of *strategic* choice: the choices of each chess player are *interdependent*: what is 'best' for one player can only be defined with reference to the choices, plans, and beliefs of the other.

Given that, the connection between the theory of games and economic behaviour should now be clear. Although economic situations are not *games* in the ordinary sense of the term, the multiple interdependencies between buyers and sellers, producers and consumers, and so on, give rise to problems of strategic choice. The best thing for, say, an automobile manufacturer to do depends upon the future demand for their automobiles, which depends on what consumers will want. And what consumers will want depends on what other automobile manufactures may produce, or even on the availability of alternative transportation such as trams or ride-sharing applications that reduce the benefit of owning an automobile. Economic behaviour, on this view, is nothing more than behaviour in a game-theoretic context, given a suitably enlarged conception of what a "game" consists of.

Game theory is the mathematical study of problems of strategic choice. It originated in 1921 with the work of the French mathematician Émile Borel, who analysed the game of poker in order to answer the question of when one should bluff. (As serious poker players know, in order to play poker well you have to bluff.) But, even though his early work was really just at the inception of game theory, Borel had a vision of its potential applications, foreseeing how it could be applied to fields of enquiry far removed from simple parlour games. Despite Borel's early efforts, the first major theoretical result was due to von Neumann (1928), who proved the influential "Minimax theorem" concerning a special class of two-player games where one player's gain is another player's loss. (Such games are known as *zero-sum* games, due to this fact about their payoff structure.) What the Minimax theorem says is that such games always have an optimal course of action for each player such that, if each player follows their respective course of action, they successfully *minimise* the *maximum* loss they might incur. Von Neumann's result was a watershed moment in the development of game theory because it showed that two-player zero-sum games had an effective "solution" guiding the outcome of play. And, given that, the next obvious question to ask was whether the same result, or a similar result, could be shown to hold for other types of games. And, to answer that, we first need to get a bit more precise about the fundamental concepts we have been talking about: what, exactly, we mean by a *game* and a *strategy*.

1.3 Game Theoretic Fundamentals

To begin, let us define a *game* as an interaction between a *finite* and *fixed* number of players, typically denoted by N. This assumption makes sense for many games of strategy like chess (two players), poker (two or more players), and bridge (four players). However, if we consider a "game" such as football, this assumption might seem less appropriate because the number of players on the pitch can vary over time due to penalties, and the *identity* of the players on the pitch can also change over time due to substitutions. In practice this does not present a problem if we conceive of things a bit differently: the total number of players *available* does not change – it corresponds to the complete roster of the team – even though not all players may be *active* at any given time. So the requirement that a game have a finite and fixed number of players raises few practical problems, provided that some flexibility is exercised in the representation.

In a game, each player can choose from one of certain number of *actions*. In the simplest games, like Rock-Paper-Scissors, each player has only a single choice of action and all actions take place simultaneously. But sometimes the actions aren't performed simultaneously, like the game shown in Figure 2a. In that game, each player has a single choice to make, and only if player A chooses A_2 will player B get to perform an action. In more complex games, like Tic-Tac-Toe, the choices the player can make may *take into account the entire history of play up to that point*.[2] A *strategy* is a plan of action that specifies what choice the player will make at every possible decision node they face in the game. Even for a simple game like Tic-Tac-Toe, strategies are vastly complex things. The first player alone has $9 \cdot 7^8 = 51,883,209$ possible ways to

---

[2] So, it's not just the state of the board at the time of play that matters, but how the two people *got* to that state. In principle, a strategy could recommend two different moves for the same state of the board, if two different histories of play led to the board looking visually the same.

make their first two moves![3] For a "real" game like chess, a strategy is an almost unbelievably complex object: after the first round of play, there are 400 possible board configurations. After the second round, there are 197,281 possible board configurations. After the third round, there are 119,060,324.

Traditional game theory represents games in two different ways. One representation explicitly tracks every possible way the game can unfold by looking at each possible move available to a player during their turn, and drawing one path for every way the game could be played, from the beginning to the end. At points where a player has a choice to make as to what to do, the path will split according to how many options the player has. This results in a structure known as a *game tree*, because every choice point for a player leads to a "branching" of possibilities (except for the last move of the game), as shown in the simple game of Figure 2a. This is known as the *extensive form representation* of a game. Another representation shows the game as a matrix, with the strategies for each player positioned along one axis and the outcome of the game denoted in the corresponding cell. This is known as the *strategic form representation*, but is also called the *normal form representation*, after von Neumann and Morgenstern, who believed that normally one could adopt this representation without any loss of generality in the subsequent analysis. The strategic form representation is perhaps most natural for games consisting of a single simultaneous move made by each player, like that of Rock-Paper-Scissors shown in Figure 2b, but it turns out that *any* extensive form game can be represented this

---

[3] This assumes all moves are independent and we do not require consistency for board positions which are strategically equivalent. For example, one strategy for X could suggest the following sequence of opening moves:

$$\begin{array}{|c|c|}\hline X & \\\hline & \\\hline & \\\hline\end{array} \rightarrow \begin{array}{|c|c|}\hline X & O \\\hline & \\\hline & \\\hline\end{array} \rightarrow \begin{array}{|c|c|}\hline X & O \\\hline & X \\\hline & \\\hline\end{array}$$

as well as this sequence of opening moves:

$$\begin{array}{|c|c|}\hline X & \\\hline & \\\hline & \\\hline\end{array} \rightarrow \begin{array}{|c|c|}\hline X & \\\hline O & \\\hline & \\\hline\end{array} \rightarrow \begin{array}{|c|c|}\hline X & \\\hline O & \\\hline X & \\\hline\end{array}.$$

This strategy is peculiar in that it responds differently to O's selection of a center edge position depending on whether it appears on the top edge or on the left edge of the board. Since these two moves, by O, are strategically equivalent (they correspond to a reflection of the board along the upper-left to lower-right diagonal, which doesn't fundamentally change the future opportunities available to O), the fact that the strategy recommends two strategically different moves suggests a kind of internal inconsistency to the strategy. Nevertheless, strictly speaking, it is permitted according to the given definition of a 'strategy'.

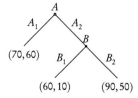

(a) An extensive form representation of a game. Player A moves first, and if she chooses strategy A_2, then player B gets to move. Payoffs are shown with A's payoff listed first.

		\multicolumn{3}{c}{Column}		
		Rock	Paper	Scissors
	Rock	(0,0)	(−1,1)	(1,−1)
Row	Paper	(1,−1)	(0,0)	(−1,1)
	Scissors	(−1,1)	(1,−1)	(0,0)

(b) The matrix representation of Rock-Paper-Scissors. The payoffs are listed in the form (Row, Column).

Figure 2 Two games illustrating the two different ways they can be represented.

way with a little bit of work.[4] In this Element, most of the time we will consider games shown in the matrix form, but in Section 4 we will consider one extensive form game known as the centipede game.

In game theory, a *solution concept* identifies the strategies which satisfy certain principles of rationality and can arguably be defended as providing a reasonable 'solution' to the game. There are many solution concepts available, but the most important one is the *Nash equilibrium*. A Nash equilibrium occurs when each individual player has settled upon a choice of strategy such that the overall collection of strategies has a certain best-reply property. Suppose that the game under consideration has N players. When each player has settled upon a strategy, the list of strategies used by each player, $\langle S_1, S_2, \ldots S_N \rangle$, is called a *strategy profile*. At a Nash equilibrium, no player has an incentive to change their strategy, provided that everyone else continues to follow the strategy allocated to them by the profile.[5]

---

[4] One simply needs to encode strategies in the right way. If a player has the possibility of making decisions at N nodes in the game tree, then a strategy for that player is a composite object $s_1 s_2 \ldots s_N$ where s_i represents the decision that player will take at node i in the game tree. The strategic form representation can then list all possible strategies for a player down the rows (or across the columns), and the payoff appearing in the corresponding cell will be the payoff realised from the path traced through the game tree by the players' strategies. Notice that this representation of a player's strategy requires specifying what that player would do at nodes in the game tree *which will not be reached* in the actual course of play.

[5] There are two subtleties to bear in mind. First, saying that no player has an incentive to change their strategy does *not* mean that a player would *do worse* if they did. At a Nash equilibrium, a change in strategy may result in a player receiving the same payout that they would have

In a Nash equilibrium, each player adopts a best response to the actions of others. One of the remarkable facts about games, proved by John Nash in 1950 (see Nash, 1950b), is that every game with a finite number of players and a finite number of strategies has at least one Nash equilibrium if we allow players the option of randomising over strategies. More precisely, a Nash equilibrium is guaranteed to exist if we expand the concept of a 'strategy' to include probability distributions over the actions available to players. (You can think of this as a player choosing to create genuine uncertainty in their opponent by selecting an action by flipping a coin, or using a randomisation device; if a player's opponent is *genuinely* uncertain as to what a player will do, then the opponent's best response needs to take into account all the actions a player may take, along with the chance that the player will do that.) In this new framework, what we have been calling a 'strategy', up to now is more properly known as a *pure strategy*; probability distributions over pure strategies are known as *mixed strategies*. Nash's result says that every game with finitely many players and strategies has at least one Nash equilibrium in mixed strategies. One further conceptual shift we must make when speaking of mixed strategies is that we need to talk about a player's *expected* payoff, since there's no guarantee about what outcome will actually result when the game is played. But if we think of a mixed strategy played against another strategy S (pure or mixed) many times, then the average payout over the long run will converge to the expected payout.

In the game of Figure 2a, one Nash equilibrium has player A choosing A_2 followed by player B choosing B_2. This gives a payoff of 90 to A and 50 to B, neither of which could be improved on: had B chosen B_1, he would have received 10 rather than 50, and had A chosen A_1, she would have received 70 rather than 90. Note that, had A chosen A_1, it is true that B would have received 60 rather than 50 – an improvement – but note that this does not result from an alternative choice of strategy by B and so is compatible with the definition of a Nash equilibrium.

In the game of Figure 2b, one can readily check that neither Rock nor Paper nor Scissors can be an equilibrium strategy for either player: if Row picks a strategy and loses, there is always a winning strategy Row could switch to. And, likewise, if Row picks a strategy and wins, then there is always another strategy that Column could switch to and win. (This is the reason underlying the evolutionary cycles in the population behaviour of *Uta stansburiana*.) However, if both players pick a strategy at *random*, with rock, paper, and scissors

---

received under their original choice. Second, a Nash equilibrium only imposes a requirement for a *single* player contemplating a change. If more than one player were to change strategies at the same time, then anything could happen.

being all equally likely, one can show that no alternative strategy exists – pure or mixed – which one could switch to that would yield a better expected payoff. To see this, first let σ denote the strategy which assigns probability $\frac{1}{3}$ to each pure strategy. For convenience, we will typically adopt the convention of writing a mixed strategy like σ as $\frac{1}{3}$Rock $+ \frac{1}{3}$Scissors $+ \frac{1}{3}$Paper.[6] Now consider the expected payoff when σ plays Rock:

$$\pi(\sigma \mid \text{Rock}) = \tfrac{1}{3}\pi(\text{Rock} \mid \text{Rock}) + \tfrac{1}{3}\pi(\text{Scissors} \mid \text{Rock}) + \tfrac{1}{3}\pi(\text{Paper} \mid \text{Rock})$$
$$= \tfrac{1}{3} \cdot 0 + \tfrac{1}{3} \cdot (-1) + \tfrac{1}{3} \cdot 1$$
$$= 0.$$

The same is obviously true when σ is played against either Scissors or Paper. From this, it can be easily shown that there is no mixed strategy μ which assigns probability r to Rock, s to Scissors, and p to Paper, where $r + p + s = 1$, such that $\pi(\mu \mid \sigma) > \pi(\sigma \mid \sigma)$.[7] This means that σ is a Nash equilibrium.

This raises an interesting question: why do populations of the common side-blotched lizard not evolve to a state consisting of equal representations of all three types? Since the common side-blotched lizards are essentially playing the game Rock-Paper-Scissors, that would be analogous to the mixed-strategy Nash equilibrium of the game underlying their evolutionary situation. To see this, we can now properly begin our discussion of *evolutionary game theory*. And the first thing we will see is that the solution concept of a Nash

---

[6] Why this notation? When each player uses a mixed strategy, writing the mixed strategies in this way allows us to compute the expected outcome of the game through a convenient abuse of notation: simply treat the mixed strategies as polynomials and multiply them. For example, suppose Player 1 uses $\sigma = \frac{1}{3}$Rock$+\frac{1}{3}$Scissors$+\frac{1}{3}$Paper and Player 2 uses $\mu = \frac{2}{3}$Rock$+\frac{1}{3}$Paper. Then the expected outcome of σ played against μ is:

$$\sigma\mu = \left(\tfrac{1}{3}\underset{\text{Player 1}}{(\text{Rock})} + \tfrac{1}{3}\underset{\text{Player 1}}{(\text{Scissors})} + \tfrac{1}{3}\underset{\text{Player 1}}{(\text{Paper})}\right)\left(\tfrac{2}{3}\underset{\text{Player 2}}{(\text{Rock})} + \tfrac{1}{3}\underset{\text{Player 2}}{(\text{Paper})}\right)$$

$$= \tfrac{2}{9}\underset{\text{Player 1 \quad Player 2}}{(\text{Rock})(\text{Rock})} + \tfrac{2}{9}\underset{\text{Player 1 \quad Player 2}}{(\text{Scissors})(\text{Rock})} + \tfrac{2}{9}\underset{\text{Player 1 \quad Player 2}}{(\text{Paper})(\text{Rock})}$$

$$+ \tfrac{1}{9}\underset{\text{Player 1 \quad Player 2}}{(\text{Rock})(\text{Paper})} + \tfrac{1}{9}\underset{\text{Player 1 \quad Player 2}}{(\text{Scissors})(\text{Paper})} + \tfrac{1}{9}\underset{\text{Player 1 \quad Player 2}}{(\text{Paper})(\text{Paper})}.$$

Each 'term' of the 'polynomial' represents a possible outcome of play, and the coefficient of each 'term' is the probability that outcome will occur.

[7] Why does $r + p + s = 1$? Because those are the probabilities that a player will use one of the three pure strategies. There are no other strategies available, and a player has to do *something*, so those probabilities must add to one.

	S_1	S_2
S_1	(2,2)	(1,1)
S_2	(1,1)	(1,1)

Figure 3 The evolutionary drift game.

equilibrium – while a perfectly reasonable solution concept in many contexts – is not the right one to help us understand evolution.

2 Evolutionarily Stable Strategies

2.1 Basic Concepts

Consider the game of Figure 3. Assume that the payoffs listed in each cell of the matrix are the expected number of offspring an individual will have as a result of the interaction, and also assume that we are talking about a species where all offspring are of the same type as their parent.[8] It can be easily seen that the game of Figure 3 has two Nash equilibria in pure strategies: one where both individuals play S_1 and another where both individuals play S_2.

But now suppose that, for historical reasons, the population is in the state where everyone follows the equilibrium strategy S_2. If an S_1-mutant appears, the mutant does not suffer a fitness disadvantage with respect to the rest of the population, because in an $S_1 \bullet\!\!-\!\!\bullet S_2$ interaction the S_1-mutant still receives a payoff of 1, which is exactly what every S_2 individual in the population receives. This means that there is no selection pressure against the S_1-mutant, and so they may persist in the population. If a *second* individual appears following the S_1 strategy (either from an independent mutation or as one of the offspring of the original mutant), the payoff from an $S_1 \bullet\!\!-\!\!\bullet S_1$ interaction is twice that earned by the S_2-type. This gives the individuals following S_1 an explicit fitness advantage over those following S_2, introducing selection pressure *against* the incumbent strategy. Even if, as we are assuming, the S_2-type is the majority of the population, over time we would expect the greater reproductive success of the S_1-type to drive the S_2-type to extinction.[9]

---

[8] This assumption may raise the question of how applicable evolutionary game theory is to sexually reproducing populations, where the offspring typically feature a blend of traits of their parents. One can introduce refinements to the models which address this, but the additional complexity, at this point, is not worth it.

[9] This intuitive argument highlights the shortcomings of the Nash equilibrium solution concept for capturing the notion of evolutionary stability. Whether or not the S_2-type would actually be driven to extinction depends on details of the underlying evolutionary dynamics. Later, we will show that this happens *always* under a dynamic known as the 'replicator dynamics', but it may not happen if the evolutionary dynamics are modelled using a discrete birth-death process.

	Hawk	Dove
Hawk	$\left(\frac{V-C}{2}, \frac{V-C}{2}\right)$	$(V, 0)$
Dove	$(0, V)$	$\left(\frac{V}{2}, \frac{V}{2}\right)$

Figure 4 The Hawk–Dove game.

The underlying problem is that the Nash equilibrium solution concept only requires that a player who deviates from their equilibrium strategy must do *no better*. That is compatible, as the game of Figure 3 shows, with the deviant player doing *exactly the same*, and it is this fact which allows a mutant to establish a foothold in the population and not be driven out. In other words, the Nash equilibrium concept is too *weak* to capture the idea of evolutionary stability.

One natural idea about how to fix this problem would be to impose a more stringent requirement on the payoff received by deviant players. What if we required that any deviation from a Nash equilibrium yielded a *worse* payoff for the deviant? This is known as a *strict* Nash equilibrium.

To see that a strict Nash equilibrium is too *strong* to capture the idea of evolutionary stability, consider the game of Figure 4. This game is known as the Hawk–Dove game and was one of the first situations to be analysed using evolutionary game theory (see Maynard Smith and Price, 1973). The Hawk–Dove game represents an idealised situation of *territorial conflict*, where two individuals compete over a resource of value $V > 0$. The pure strategies available are 'Hawk', which aggressively pursues the resource and is willing to fight over it, and 'Dove', which begins with aggressive display behaviour but quickly backs down when challenged. If two Hawks meet, a fight will break out and each player has an equal chance of losing. The loser incurs a cost $C > 0$, which is why the expected fitness for a Hawk–Hawk interaction is $\frac{V-C}{2}$: each Hawk has a 50 per cent chance it will win and receive V, and a 50 per cent chance it will lose and bear the cost $-C$. When a Hawk meets a Dove, the Dove retreats in the face of escalation, and so the Hawk receives the entire value of the resource and the Dove receives nothing. When two Doves meet, they each share the resource.

If $V > C$, Hawk is a *strictly dominant* strategy, which means it is always in one's interest to play Hawk no matter what one's opponent does. Another way of stating this is that the Hawk – Hawk strategy profile is a strict Nash equilibrium. In this case, we would expect the strategy Hawk to be evolutionarily stable, since a population consisting exclusively of Hawks could not be invaded by Doves because Doves would have a strictly lower fitness. Those with a sharp eye for definitional details will spot that this suggests a strict Nash equilibrium strategy should be *sufficient* for evolutionary stability. In some cases, that is correct. (Complications will be explored in Section 4.) But what we are about to see is that it is not *necessary* to have a strict Nash equilibrium strategy in order to have evolutionary stability.

The Hawk–Dove game is interesting when the cost incurred by the loser, when a fight breaks out, is greater than the total value of the resource. When $V < C$, there is no pure strategy Nash equilibrium but there is one in mixed strategies. If σ^* denotes the strategy where each individual plays Hawk with probability $p = \frac{V}{C}$ and Dove with probability $1-p$, the use of that mixed strategy by both players is a Nash equilibrium.[10] It is easy to show that,[11] when one player uses the Nash equilibrium strategy σ^* in the Hawk–Dove game, the *other player's* payoff is such that

$$\pi(\text{Hawk} \mid \sigma^*) = \pi(\text{Dove} \mid \sigma^*) = \pi(\sigma^* \mid \sigma^*).$$

This means that, if one player uses σ^*, then for the other player *every possible* mixed strategy μ earns the same payoff when played against σ^* that σ^* earns when it plays against itself. Given that, one might ask whether σ^* is evolutionarily unstable, for isn't this essentially the same situation that we encountered in the Evolutionary Drift game of Figure 3?

Actually, no. The interesting observation made by Maynard Smith and Price was that, even though σ^* is not a strict Nash equilibrium strategy, it was impossible for evolutionary drift to occur. Why? Because evolutionary stability doesn't just depend on what happens when a single mutant *enters* a population. Evolutionary stability also depends on what happens when the number of mutants starts to increase and mutants begin to interact with their own kind. And this is really what went wrong in the game of Figure 3; there, mutants did *even better* than the incumbent strategy when interacting with their own kind.

In the case of the Hawk–Dove game, a little bit of algebra allows one to show that, for any mutant strategy $\mu \neq \sigma^*$, it is the case that the incumbent strategy σ^* has a higher expected payoff playing against the mutant than the mutant has when playing against itself. Or, in symbols, that $\pi(\sigma^* \mid \mu) > \pi(\mu \mid \mu)$. The proof of this isn't difficult and follows from a bit of algebra.[12] This shows that

---

[10] Why is there a * attached to the σ? That decoration is typically used to denote a strategy which, when played, is part of a Nash equilibrium.

[11] This is an instance of a more general result known as the fundamental theorem of mixed-strategy Nash equilibria. If you are interested in the details, Gintis (2009) is a good place to begin.

[12] Since a mixed strategy selects a pure strategy at random, calculating the expected payoffs of $\pi(\sigma^* \mid \mu)$ and $\pi(\mu \mid \mu)$ requires determining each possible pure strategy pairing between the two players, and then summing up the payoffs using the combined probabilities as weights. For example, suppose that μ plays Hawk with probability q and Dove with probability $1-q$. Then

$$\pi(\sigma^* \mid \mu) = pq \cdot \pi(\text{Hawk} \mid \text{Hawk}) + p(1-q) \cdot \pi(\text{Hawk} \mid \text{Dove})$$
$$+ (1-p)q \cdot \pi(\text{Dove} \mid \text{Hawk}) + (1-p)(1-q) \cdot \pi(\text{Dove} \mid \text{Dove}).$$

the mixed strategy σ^* would – even though it is not a strict Nash equilibrium – be evolutionarily stable because it has a higher expected payoff when played against any mutant than the *mutant* has *when played against its own kind*.

Maynard Smith and Price turned this last observation into the definition of an *evolutionarily stable strategy*, as follows:

Definition 1: Evolutionarily Stable Strategy (Original Definition)

A strategy σ is an *evolutionarily stable strategy* (ESS) if and only if for all strategies $\mu \neq \sigma$ it is the case that:

either $\pi(\sigma \mid \sigma) > \pi(\mu \mid \sigma)$,
or $\pi(\sigma \mid \sigma) = \pi(\mu \mid \sigma)$ and $\pi(\sigma \mid \mu) > \pi(\mu \mid \mu)$.

The Maynard Smith and Price definition of an ESS is slightly unusual in that it is a *disjunctive* definition with an exclusive-or. This is because, although a strict Nash equilibrium is too strong to capture *every* aspect of evolutionary stability, it is certainly true that a strict Nash equilibrium strategy will be evolutionarily stable: it just happens that there are other strategies which are evolutionarily stable as well. So the way to understand Definition 1 is as follows: first, every strategy which is a strict Nash equilibrium is an ESS. Second, if a strategy σ is not a strict Nash equilibrium, then it must be the case that σ is a *better* response to every other mutant μ than μ is to itself. An ESS strengthens the Nash equilibrium solution concept by imposing a second-order best-reply condition.

It is also common to find an ESS defined as follows:

Definition 2: Evolutionarily Stable Strategy (Alternative Definition)

A strategy σ is an *evolutionarily stable strategy* (ESS) if and only if for all strategies $\mu \neq \sigma$,

1. $\pi(\sigma \mid \sigma) \geq \pi(\mu \mid \sigma)$.
2. If $\pi(\sigma \mid \sigma) = \pi(\mu \mid \sigma)$, then $\pi(\sigma \mid \mu) > \pi(\mu \mid \mu)$.

Definition 2 is, of course, logically equivalent to Definition 1. But what Definition 2 makes clear is that every ESS is a Nash equilibrium (condition one), a fact which the original disjunctive definition obscures. It also makes clear that an ESS is a strengthening of the Nash equilibrium concept, as condition

---

Plugging in the payoffs from Figure 4 and using the fact that $p = \frac{V}{C}$ allows one to show the relative ordering between $\pi(\sigma^* \mid \mu)$ and $\pi(\mu \mid \mu)$.

two imposes a further constraint on the payoffs in the case that σ is not a strict Nash equilibrium. In the Evolutionary Drift game, although strategy S_2 satisfies condition one it fails to satisfy condition two; in the Hawk–Dove game, when $V < C$ the mixed strategy Nash equilibrium satisfies *both* conditions.

At this point, we have used the Hawk–Dove game and some facts about Nash equilibria to provide intuitive motivation for adopting the definition of an ESS mentioned above (in either logical form). One might well ask why *that* particular definition is worth using. How do we know that there aren't other alternative conceptions of evolutionary stability, each with their own intuitive support, which turn out to be different but equally worthwhile? That's a fair question and a point worth exploring. As we will see, there are a number of conceptual subtleties in play here.

One intuition regarding evolutionary stability is that any trait, behaviour, or strategy worthy of the label needs to be strictly better than those competitors which are, in some sense, 'close' to it.[13] Since mutation tends to introduce small modifications to the underlying DNA,[14] it is unlikely that the difference between an individual and their offspring will be so radical as to flip completely between one behaviour and a hugely different variant in a single mutation. Even though we are currently working at a very high level of description, not worrying about how strategies are encoded genetically, or about the fine-grain details of mutational change, a rough first approximation can be developed as follows. Let σ and μ be two strategies with $\sigma \neq \mu$. If $\epsilon > 0$ is sufficiently small, the mixed strategy, $\tau = \epsilon\mu + (1-\epsilon)\sigma$, is 'close' to σ in the sense that it isn't that great a deviation from σ, regardless of the nature of μ. If you imagine τ being used many times in a repeated game, the outcome of playing τ will mostly look the same as playing σ. For games with at least two pure strategies, we can meaningfully speak of all of the strategies which fall within distance ϵ from any particular strategy σ. (This is known as a "ϵ-neighbourhood" of σ.)[15] At this

---

[13] This is not the same thing as being a strict Nash equilibrium. A strict Nash equilibrium requires that *every* possible change in strategy results in a lower payoff. Here we are exploring the idea that only changes to 'nearby' strategies result in lower payoffs.

[14] Not all mutation results in small modifications to the underlying DNA. During meiosis, chromosomal pairs are matched before separating during cell division and, occasionally, sections of DNA from the paired chromosomes can be exchanged. This results in differences much larger than the 'point mutations' which only affect a single base pair. Be that as it may, we are only thinking about the degree of change possible via mutation as an *intuition pump*, to help clarify what we mean when we talk about evolutionary stability. In our discussion of evolutionarily stable strategies, we are abstracting entirely from the question of how a strategy is encoded at the underlying 'genetic' level and how mutation *actually* leads to strategic variation.

[15] Those with a mathematical background will recognise that a number of measure-theoretic subtleties are being dealt with rather casually. For those interested in the details, one excellent source is Weibull (1995).

point, we have the necessary conceptual tools to state the following definition, first due to Hofbauer, Schuster, and Sigmund (1979).

Definition 3: Local Superiority

A strategy σ is said to be *locally superior* if and only if there is some ϵ-neighbourhood of σ such that, for every strategy $\mu \neq \sigma$ in that neighbourhood, $\pi(\sigma \mid \mu) > \pi(\mu \mid \mu)$.

It can be proven (see the abovementioned paper for details) that a strategy is an ESS if and only if it is locally superior. This is slightly surprising because, comparing the form of Definitions 1 and 3, it looks as though local superiority is a slightly stronger condition. The requirement that $\pi(\sigma \mid \mu) > \pi(\mu \mid \mu)$, for example, only appears in the second disjunct of Definition 1. Nevertheless, they turn out to be logically equivalent. This fact gives us additional reason to believe that the Maynard Smith and Price definition captured something important.

A second intuition about evolutionary stability is that such strategies should be resilient in the face of invading mutants. In the Evolutionary Drift game, S_2 had no resilience at all. But there are limits to how much resilience could be reasonably required. Consider the strategy S_1 in the game of Figure 5. That strategy satisfies the definition of an ESS, and hence is also locally superior. Now imagine we have a large population of individuals, all of whom interact in pairs and play the game. Assume there is no overlapping of generations, so that after each round of play the current population is replaced by their offspring. If there is some chance of mutation occurring, although it is extremely unlikely that, between one generation and the next, the *vast majority* will experience the exact same mutation which causes them to follow strategy S_2 instead of S_1, such an unlikely event is still possible. Yet the mere *possibility* of a freak occurrence should not be seen as challenging the claim that S_1 is evolutionarily stable. It wasn't, so to speak, S_1's fault that it was eliminated by almost every single offspring spontaneously mutating to another type: that was simply the

	S_1	S_2
S_1	$(1,1)$	$(0,0)$
S_2	$(0,0)$	$(1,1)$

Figure 5 The two-ways-of-life game, also known as a perfect coordination game. One example of a real-world problem having this structure is deciding what side of the road to drive on: it doesn't really matter whether people choose to drive on the left- or right-hand side of the road, as long as they all do the same thing.

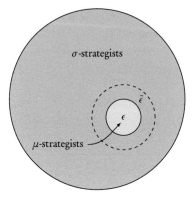

Figure 6 A representation of the population, with both σ- and μ-strategists shown, as used in the definition of a uniform invasion barrier.

rare product of chance. What this thought experiment shows is that that there are limits on how resilient we can reasonably require a strategy to be, because otherwise nothing would end up being called evolutionarily stable.

One way to capture this idea of 'reasonable robustness' in the face of mutation is found in another definition of evolutionary stability, also due to Hofbauer, Schuster, and Sigmund (1979). Consider a population where, initially, everyone follows the strategy σ, and suppose that some small proportion $0 < \epsilon \ll 1$ of mutants following the strategy μ appear. Here the thought is that, rather than considering the case of a *single* mutant, we are considering the case where a small cluster of mutants simultaneously appear, as illustrated in Figure 6. If all pairwise interactions occur at random, and all members of the population are equally likely to be chosen for an interaction, then there is an ϵ chance that someone will interact with a μ-strategist and a $1-\epsilon$ chance of interacting with a σ-strategist. A population is said to have a *uniform invasion barrier* if it is the case that there is some fixed threshold such that, as long as the number of mutants is *below* that threshold, then the expected fitness of the incumbent strategy σ is higher than that of the mutants. Definition 4 provides a precise mathematical statement of this idea.

Definition 4: Uniform Invasion Barrier

A strategy σ is said to have a *uniform invasion barrier* if and only if there exists an $\bar{\epsilon} > 0$ such that for all strategies $\mu \neq \sigma$ and every $\epsilon < \bar{\epsilon}$,

$$\epsilon \cdot \pi(\sigma \mid \mu) + (1-\epsilon) \cdot \pi(\sigma \mid \sigma) > \epsilon \cdot \pi(\mu \mid \mu) + (1-\epsilon) \cdot \pi(\mu \mid \sigma).$$

Although Definitions 1 and 4 appear on the surface to be quite different – the first concerning a strengthening of the Nash equilibrium solution concept, the second concerning the relative expected fitness of strategies in a mixed population – they *also* turn out to be equivalent. A strategy is an ESS if and only if it has a uniform invasion barrier. The reason for this surprising equivalence is that, from a mathematical point of view, there is no difference in your expected payoff if your opponent is selected at random from a mixed population containing ϵ type S_1 individuals and $1-\epsilon$ type S_2 individuals, or if your opponent uses the mixed strategy $\epsilon \cdot S_1 + (1-\epsilon) \cdot S_2$.

Putting all these results together, we then have that a strategy is an ESS if and only if it is locally superior if and only if it has a uniform invasion barrier. The fact that three ostensibly different concepts of evolutionary stability turn out to be logically equivalent is evidence that the definition of an ESS isn't entirely arbitrary but rather captures a core notion of evolutionary stability.[16]

2.2 Properties of an ESS

One of the great results from traditional game theory is that every game with finitely many players and a finite number of strategies has at least one Nash equilibrium. An ESS strengthens the Nash equilibrium concept to such an extent that this result does not hold for it. To see this, consider again the game of Rock-Paper-Scissors from Figure 2b. This game has exactly one Nash equilibrium: the strategy σ which selects Rock, Paper, or Scissors at random, assigning equal probability to all three pure strategies. It is easy to check that $\pi(\sigma \mid \sigma) = \pi(\text{Rock} \mid \sigma) = 0$. When σ plays itself, all of the nine possible pure strategy pairings happen with equal probability, giving an expected payoff of 0. When Rock plays σ, there's a one-third chance σ will choose Rock and tie, a one-third chance it will select Paper and win, and a one-third chance it will select Scissors and lose. According to Definition 1, we now need to check whether the second disjunct is satisfied. Since $\pi(\sigma \mid \text{Rock}) = 0$ and $\pi(\text{Rock} \mid \text{Rock}) = 0$, it follows that the second disjunct is not satisfied and so σ is not an ESS. Because it is a necessary condition that an ESS is a Nash equilibrium, this shows that the game of Rock-Paper-Scissors has no ESS. This proves the following theorem.

Theorem 1 *A game is not guaranteed to have an ESS.*

---

[16] The situation is not unlike what we find in theoretical computer science regarding the concept of a computable function. Because one can prove that the class of functions capable of being calculated using a Turing machine, or a cellular automaton, or recursive functions (or even quantum computers!) are all *exactly* the same, that gives us reason to believe that we have successfully characterised the idea of a computable function.

	A	B
A	$(1,1)$	$(100,0)$
B	$(0,100)$	$(100,100)$

Figure 7 A game with two Nash equilibria in pure strategies, one of which is weakly dominated.

Another interesting consequence of the definition of an ESS can be seen in the game of Figure 7. This game has exactly two Nash equilibria, both in pure strategies: (A,A) and (B,B).[17] But take a close look at the (B,B) equilibrium. Because $\pi(B \mid B) = 100 = \pi(A \mid B)$, we need to check the second clause of Definition 1. Since $\pi(B \mid A) = 0 < \pi(A \mid A) = 1$, it follows that B is not an ESS. It is easily seen that (A,A) is a strict Nash equilibrium, because $\pi(A \mid A) > \pi(B \mid A)$, which satisfies the first clause of Definition 1, and so A is an ESS. This is surprising because the payoff at the (B,B) equilibrium is far superior to the payoff of the (A,A) equilibrium!

This last example illustrates a more general phenomenon. In traditional game theory, it is common to talk about *strictly* and *weakly* dominated strategies. (See Definition 5.) Since no strictly dominated strategy is ever part of a Nash equilibrium, it follows that a strictly dominated strategy cannot be an ESS. But weakly dominated strategies *can* appear in a Nash equilibrium, and even result in the best possible payoffs for all players. Nevertheless, one can prove a weakly dominated strategy precludes a strategy from being an ESS.

Definition 5: Strictly and Weakly Dominated Strategies

A strategy σ is *strictly dominated* if and only if there exists another strategy μ which guarantees the player a higher payoff than σ, regardless of the strategies adopted by the other players.

A strategy σ is *weakly dominated* if and only if there exists another strategy μ such that σ does worse than μ, in some cases, against the opposing players while, in the remaining cases, σ does no better than μ.

Theorem 2 *No weakly dominated strategy is an ESS.*

Now consider the game of Figure 8. This game has a surprising number of Nash equilibria: three in mixed strategies and four in pure strategies. Consider the (A,A) equilibrium. Inspection of the payoff matrix shows that, faced with

---

[17] There is no mixed-strategy Nash equilibrium, because if your opponent plays a mixed strategy $pA + (1-p)B$, for any $0 \leq p \leq 1$, your best response is to play the pure strategy A.

	A	B	C
A	(3,3)	(0,0)	(0,0)
B	(0,0)	(1,1)	(1,2)
C	(0,0)	(2,1)	(1,1)

Figure 8 A game with multiple ESSes. The four Nash equilibria in pure strategies are shaded. The two which correspond to ESSes are indicated by a darker shade of gray.

an opponent playing *A*, all other strategies except *A* do strictly worse, receiving payoffs of 0 instead of a payoff of 3. That means strategy *A* satisfies the first clause of Definition 1 and so is an ESS. But now consider the (*C*, *C*) equilibrium, which is more interesting. When faced with an opponent playing *C*, strategy *A* is clearly worse (receiving a payoff of 0 instead of 1) but strategy *B* does just as well when played against *C* as *C* does! Does this mean that *C* is not an ESS? No, because when a payoff tie like this occurs, we have to check whether the second-order best-reply condition is met. In this case, we find that, when faced with an opponent playing strategy *B*, that strategy *C* does strictly better against *B* than *B* does when played against itself. In symbols: $\pi(B \mid C) > \pi(C \mid C)$, because $2 > 1$. That means that strategy *C* is *also* an ESS.

The game of Figure 8 illustrates two important facts about ESS. First, games may have more than one ESS. Although the ESS solution concept is a strengthening of the Nash equilibrium solution concept – so much so that, unlike a Nash equilibrium, an ESS is not guaranteed to exist – it is not strong enough to guarantee uniqueness when we know at least one ESS does exist. Second, when games have multiple ESSes, it may be the case that the payoff received by one ESS is strictly worse than the other. This possibility nicely reflects what we find in real life: evolution does not always produce optimal outcomes, as Stephen J. Gould observed in his discussion of the panda's thumb.[18] Evolution may lead to a species being 'stuck' at an outcome which is evolutionarily stable yet not optimal.[19]

Now that we know that, when an ESS exists, it may not be unique, the next question to ask is *how many* ESSes can there be for a given game? When it comes to Nash equilibria, even a very simple game might have infinitely

---

[18] See Gould (1980) for the relevant essay.

[19] A cultural evolutionary example may be provided by the layout of the standard QWERTY typewriter. This particular keyboard layout was deliberately designed to *slow down* the typist, so as to reduce the chance of two keys getting stuck. The success of that particular model, combined with the cost of having to retrain on a different layout, led to its becoming the standard layout throughout the English-speaking world. (However, there are minor differences in the positioning of certain less-used keys between US English and UK English layouts.)

	Do This	Do That
Do This	(0,0)	(0,0)
Do That	(0,0)	(0,0)

The Game of Life

Figure 9 The fundamental existentialist dilemma, represented as a game.

many Nash equilibria. For the 'game of life', shown in Figure 9, *every* strategy, both pure and mixed, is a Nash equilibrium strategy. A bit of reflection on the game of life will soon reveal that the reason every strategy is a Nash equilibrium strategy is because they all have the exact same payoff: zero. If strategies overlap, differing only in the probability assigned to a choice, like $\sigma = \frac{1}{3}$Do This $+ \frac{2}{3}$Do That and $\mu = \frac{2}{3}$Do This $+ \frac{1}{3}$Do That, it makes no material difference to the resulting payoffs.

However, when it comes to evolutionarily stable strategies, the strengthening of the Nash equilibrium solution concept means that two ESSes cannot overlap in the way that was possible in the game of life. To make this precise, we first need to introduce the idea of the *support* of a mixed strategy.

Definition 6: The Support of a Mixed Strategy

Let $S = \{ S_1, \ldots, S_n \}$ be a set of pure strategies and let $\sigma = p_1 S_1 + \cdots + p_n S_n$ be a mixed strategy over S. Then the *support* of σ, denoted supp(σ), is the set of all pure strategies which σ assigns a nonzero probability of being played.

From this, one can prove the following:

Theorem 3 *Let σ be an evolutionarily stable strategy (ESS) for a game G. If μ is a strategy and* supp(μ) $\subset$ supp(σ), *then μ is not an ESS. If μ is an ESS and* supp(μ) = supp(σ), *then $\mu = \sigma$.*

For a detailed proof, see Alexander (2016). The reason the theorem holds is that the second-order best-reply condition required of an ESS prevents one ESS from being 'contained' within another.

Since any game G with strategy set $S = \{ S_1, \ldots, S_n \}$ has exactly $2^n - 1$ possible support sets,[20] it immediately follows that:

Theorem 4 *The number of ESSes for a game is finite.*

---

[20] The set S has 2^n subsets, but one of the subsets is the empty set. As every strategy must assign nonzero probability to at least one strategy in S, the empty set is not a possible support set.

And another consequence of Theorem 3 is that we can state a condition which allows us to identify when an ESS is unique:

Theorem 5 *A completely mixed ESS is the only ESS of the game.*

2.3 Evolutionarily Stable Sets

At this point, our analysis of evolutionarily stability has mostly concentrated on the Maynard Smith and Price concept of an evolutionarily stable strategy, which only applies to a *single* strategy. To the extent that we have examined other competing concepts, like that of local superiority or uniform invasion thresholds, we have done so only to show that they do not succeed in picking out a separate family of strategies from an ESS, but rather provide an alternative way of characterising ESSes. But these alternative characterisations still only apply to single strategies.

Yet it is possible to find games which suggest that the single-strategy focus of an ESS does not exhaust everything there is to say about evolutionary stability in games. Consider the game of Figure 10a. This game has no evolutionarily stable strategy. If we look closely at what is going on, we find something interesting. First, consider the two-strategy subgames shown in Figures 10b and 10c. Each of these subgames was formed by deleting one strategy from the original three-strategy game, and *both* two-strategy subgames have an ESS! In the case of Subgame 1, the mixed strategy ESS assigns equal probability to S_1 and S_2 and, similarly, in the case of Subgame 2, the mixed strategy ESS assigns equal probability to S_2 and S_3. This suggests that the reason the original

	S_1	S_2	S_3
S_1	$(1,1)$	$(2,3)$	$(2,2)$
S_2	$(3,2)$	$(0,0)$	$(4,1)$
S_3	$(2,2)$	$(1,4)$	$(3,3)$

(a) A three-strategy game with no ESS.

	S_1	S_2
S_1	$(1,1)$	$(2,3)$
S_2	$(3,2)$	$(0,0)$

Subgame 1

(b) A two-strategy subgame for which $\sigma = \frac{1}{2}S_1 + \frac{1}{2}S_2$ is an ESS.

	S_2	S_3
S_2	$(0,0)$	$(4,1)$
S_3	$(1,4)$	$(3,3)$

Subgame 2

(c) Another two-strategy subgame for which $\tau = \frac{1}{2}S_2 + \frac{1}{2}S_3$ is an ESS.

Figure 10 The game used by Thomas (1985) to motivate the definition of an evolutionarily stable *set* of strategies.

three-strategy game *doesn't* have an ESS might be due to a curious interaction going on between the various possible strategies. Let's take a closer look.

Why isn't σ an ESS of the original game? First, it *is* true that both σ and τ are *Nash equilibria* of the original game. What happens, though, is that neither one satisfies the stronger condition required by an ESS *on its own*. Consider the potential invading mutant $\mu_1 = \frac{3}{8}S_1 + \frac{3}{8}S_2 + \frac{1}{4}S_3$. (Subscripts are attached to the mutant because we are going to be looking at several examples involving different mutants and it will be useful to be able to distinguish between them.) Straightforward calculation shows that $\pi(\sigma \mid \sigma) = \frac{3}{2}$ and that $\pi(\mu_1 \mid \sigma) = \frac{3}{2}$, which according to either definition of an ESS requires checking whether the second-order best-reply condition holds. When we do, we find that $\pi(\sigma \mid \mu_1) = \frac{15}{8}$ and that $\pi(\mu_1 \mid \mu_1) = \frac{15}{8}$, so σ isn't an ESS! But now look at what happens when τ interacts with μ_1. First, we find that $\pi(\tau \mid \tau) = 2 = \pi(\mu_1 \mid \tau)$, so once again we have to check whether the second-order best-reply condition holds. When we do, calculation shows that $\pi(\tau \mid \mu_1) = 2$, which *is greater than* the value we just calculated for $\pi(\mu_1 \mid \mu_1)$! So, although σ doesn't satisfy the definition of an ESS with respect to μ_1, it is the case that τ does. Intuitively this suggests that although μ_1 would be able to invade a population consisting purely of σ-players, if there are some τ-players around they would be able to fight off the μ-mutants.

So why isn't τ an ESS? Because we can find another mutant strategy μ_2 which τ isn't able to fight off but σ can! Consider $\mu_2 = \frac{1}{8}S_1 + \frac{5}{8}S_2 + \frac{1}{4}S_3$. The relevant calculations to check are the following:

$$\pi(\tau \mid \tau) = 2 = \pi(\mu_2 \mid \tau) \qquad \text{and} \qquad \pi(\tau \mid \mu_2) = \frac{3}{2} = \pi(\mu_2 \mid \mu_2),$$

so τ isn't an ESS, but then we find

$$\pi(\sigma \mid \sigma) = \frac{3}{2} = \pi(\mu_2 \mid \sigma) \qquad \text{but} \qquad \pi(\sigma \mid \mu_2) = \frac{13}{8} > \frac{3}{2} = \pi(\mu_2 \mid \mu_2).$$

Although a population consisting purely of τ-players could not drive out μ_2, if there are any σ-players present they would be able to drive out the mutant.

This suggests that although *neither σ nor τ are* evolutionarily stable strategies in their own right, the *collection* $\{\sigma, \tau\}$ might have its own sense of 'evolutionary stability' which we need to identify and formalise in a definition which treats evolutionary stability as a property of *collections* of strategies. This is a step in the right direction, but the full story is a little more subtle.

Is it *always* true, for any mutant μ_3 differing from both σ and τ, that *either* σ or τ would satisfy the definition of an ESS, with respect to μ_3, as we saw earlier? Somewhat surprisingly, the answer is *no*. Consider the mutant $\mu_3 = \frac{1}{4}S_1 + \frac{1}{2}S_2 + \frac{1}{4}S_3$. Checking the relevant calculations reveals

$$\pi(\sigma \mid \sigma) = \tfrac{3}{2} = \pi(\mu_3 \mid \sigma) \qquad \text{and} \qquad \pi(\sigma \mid \mu_3) = \tfrac{7}{4} = \pi(\mu_3 \mid \mu_3)$$
$$\pi(\tau \mid \tau) = 2 = \pi(\mu_3 \mid \tau) \qquad \text{and} \qquad \pi(\tau \mid \mu_3) = \tfrac{7}{4} = \pi(\mu_3 \mid \mu_3)$$

which shows that *neither* σ nor τ satisfy the ESS definition, with respect to μ_3. And it turns out that there are actually *infinitely* many potential mutant strategies μ_i which, like μ_3, would not have either σ or τ satisfying the definition of an ESS with respect to μ_i. Here are some of them:

$$\mu_4 = \tfrac{1}{16}S_1 + \tfrac{1}{2}S_2 + \tfrac{7}{16}S_3$$
$$\mu_5 = \tfrac{1}{8}S_1 + \tfrac{1}{2}S_2 + \tfrac{3}{8}S_3$$
$$\mu_6 = \tfrac{1}{4}S_1 + \tfrac{1}{2}S_2 + \tfrac{1}{4}S_3$$
$$\mu_7 = \tfrac{3}{8}S_1 + \tfrac{1}{2}S_2 + \tfrac{1}{8}S_3$$
$$\mu_8 = \tfrac{7}{16}S_1 + \tfrac{1}{2}S_2 + \tfrac{1}{16}S_3$$

$$\vdots$$

Note that all of the μ_i strategies just listed have a curious property: they are linear combinations of σ and τ. For example, $\mu_7 = \tfrac{3}{4}\sigma + \tfrac{1}{4}\tau$. (That's why all of them assign probability $\tfrac{1}{2}$ to S_2.) For ease of future reference, let's use L to denote the set of all mixed strategies which can be written as a linear combination $q \cdot \sigma + (1-q) \cdot \tau$ for some $0 \le q \le 1$.

If you go back and check the form of μ_1 and μ_2 at the start of this subsection – the two mutant strategies for which one of σ or τ *did* satisfy the ESS definition – you will find that neither of those can be written as a linear combination of σ and τ. Moreover, if we check the ESS definition for μ_7 with respect to μ_1 – one of the mutants which cannot be written as a linear combination of σ and τ – we find that $\pi(\mu_7 \mid \mu_7) = \pi(\mu_1 \mid \mu_7)$ and $\pi(\mu_7 \mid \mu_1) > \pi(\mu_1 \mid \mu_1)$. So even though μ_7 is not a Nash equilibrium of the original game, it nevertheless satisfies the definition of an ESS with respect to μ_1!

Putting all this together, what we find with respect to the three-strategy game of Figure 10a is as follows. Although the game has no single ESS, the set L consisting of all linear combinations of the two mixed-strategies σ and τ has two properties:

1. For any strategy μ which is not a member of L, there is some strategy σ in L that satisfies the ESS definition with respect to μ. (Strictly speaking, because L contains infinitely many strategies, there will actually be *infinitely* many strategies which satisfy the ESS definition with respect to μ, even if not *every* strategy in L does.)

2. For any pair of strategies σ and τ in L, neither strategy will satisfy the definition of an ESS with respect to the other.

All of this is to motivate the following definition of an *evolutionarily stable set*, first proposed by Thomas (1985).

Definition 7: Evolutionarily Stable Set (ES Set)

Let L be a closed, non-empty set of strategies for some game G. Then L is an *evolutionarily stable set* if and only if for every strategy $\sigma \in L$ there is an ϵ-neighbourhood N around σ such that, for every strategy $\mu \in N$, it is the case that $\pi(\sigma \mid \sigma) = \pi(\mu \mid \sigma)$ and

1. If $\mu \in L$, then $\pi(\sigma \mid \mu) = \pi(\mu \mid \mu)$. (That is, σ does not satisfy the definition of an ESS with respect to μ.)
2. If $\mu \notin L$, then $\pi(\sigma \mid \mu) > \pi(\mu \mid \mu)$. (That is, σ does satisfy the definition of an ESS with respect to μ.)

The concept of an evolutionarily stable set provides one natural generalisation of the idea of an evolutionarily stable strategy. A nice property of ES sets is that, if the set L consists of a single strategy (a set consisting of a single item is an extreme example of closed set), then that strategy is an ESS. The definition of an ES set is a weakening of the concept of an ESS.

Is that the end of the story? Not even close. There are other set-based definitions of evolutionary stability that have been proposed, each of which attempt to isolate some particular intuition underlying the idea of 'stability'. For example, Balkenborg and Schlag (2001) distinguish between (i) *simple evolutionarily stable sets*, (ii) *pointwise uniform evolutionarily stable sets*, and (iii) *uniform evolutionarily stable sets*. These three definitions provide three different characterisations, of varying strength, of how a set of strategies can count as 'stable'.

For simplicity, let us consider a case where there are only two possible strategies, σ and μ, and let us suppose that the population begins in a state where everyone follows σ. Following Balkenborg and Schlag, we shall adopt the following terminology: if a small fraction ϵ of the population switches to μ *and* it is the case that the expected fitness of σ is greater than the expected fitness of μ, then we shall say that μ *is ϵ-driven out by σ*. If, on the other hand, the expected fitness of μ is greater than the expected fitness of σ, then we shall say that μ *spreads given σ*.

One of the interesting points about the Balkenborg and Schlag approach is that instead of thinking about stability from the point of view of payoffs earned by particular pairings of strategies – which is what the standard ESS approach does – they think about stability in terms of the *expected payoff of a strategy in a mixed population*. In what follows, we will use $W(\sigma)$ to denote the expected fitness of σ in a population.[21]

We can now give precise characterisations of the three set-based notions of evolutionary stability by Balkenborg and Schlag. To begin, consider the following:

Definition 8: Simple Evolutionarily Stable Set

Let L be a non-empty set of strategies for some game G. Then L is a *simple evolutionarily stable set* if and only if the following two conditions hold:

1. No strategy in L can spread, given any other strategy in L.
2. Every strategy in L can ϵ-drive out every strategy $\mu \notin L$, for some $\epsilon_\mu > 0$.

A simple evolutionarily stable set is the weakest of the three notions of evolutionary stability just mentioned. The reason for this is that the ϵ-threshold for which strategies in L can drive out invading mutants is allowed to vary according to the mutant. To see how and why this can make a difference, contrast this definition with the following one.

Definition 9: Pointwise Uniform Evolutionarily Stable Set

Let L be a non-empty set of strategies for some game G. Then L is a *pointwise uniform evolutionarily stable set* if and only if

1. No strategy in L can spread, given any other strategy in L.
2. For every strategy σ in L, there exists an $\bar{\epsilon} > 0$ such that σ will $\bar{\epsilon}$-drive out any mutant strategy not in L.

It is possible for a set of strategies to be a simple evolutionarily stable set but not a pointwise uniform evolutionarily stable set. To see this, consider the following game, first discussed by Vickers and Cannings (1987). Suppose that there are infinitely many strategies $S = \{S_1, S_2, S_3, \dots\}$ available to the players,

---

[21] Strictly speaking, the expected fitness of a strategy in a population depends on the particular composition of the population, and that dependency is not reflected in the $W(\cdot)$ notation. However, in the subsequent discussions the particular composition of the population will be clear from the context.

with the payoff function $\pi(S_i \mid S_j)$ as shown in matrix M, where S_i determines the *row* of the payoff matrix, and S_j determines the *column*.

$$M = \begin{pmatrix} 0 & -1 & -1 & -1 & \cdots \\ -\frac{1}{2} & 0 & -\frac{1}{2} & -\frac{1}{2} & \cdots \\ -\frac{1}{3} & -\frac{1}{3} & 0 & -\frac{1}{3} & \cdots \\ -\frac{1}{4} & -\frac{1}{4} & -\frac{1}{4} & 0 & \cdots \\ \vdots & \vdots & \vdots & & \ddots \end{pmatrix}.$$

That is, the payoff function is defined according to this rule:

$$\pi(S_i \mid S_j) = \begin{cases} 0 & \text{if } i = j, \\ -\frac{1}{i} & \text{if } i \neq j. \end{cases}$$

Now consider the set $L = \{ S_1 \}$ and assume that players can only choose pure strategies. The set L is a simple evolutionarily stable set because condition (1) is trivially satisfied, and condition (2) is satisfied because, for any mutant S_k not in L, we can find a sufficiently small threshold ϵ_k such that as long as the proportion of invading S_k-mutants is less than ϵ_k, the incumbent strategy S_1 has a greater expected fitness. For example, if ϵ of the population follows S_k, then the expected fitness calculations for S_1 and S_k are

$$W(S_1) = (1-\epsilon) \cdot \pi(S_1 \mid S_1) + \epsilon \cdot \pi(S_1 \mid S_k)$$
$$= -\epsilon$$

and

$$W(S_k) = (1-\epsilon) \cdot \pi(S_k \mid S_1) + \epsilon \cdot \pi(S_k \mid S_k)$$
$$= -\frac{1-\epsilon}{k}.$$

When is $W(S_1) > W(S_k)$? If we take $\epsilon_k = \frac{1}{k+1}$, then S_1 is able to drive out S_k for any $\epsilon < \epsilon_k$. Hence, $L = \{ S_1 \}$ is a simple evolutionarily stable set.

However, this also serves to show why L is *not* a pointwise uniform evolutionarily stable set. Pointwise uniformity imposes the stricter requirement that, for every strategy in L, we must be able to find a *single* threshold such that, as long as any invading mutant does not exceed that proportion of the population, the incumbent strategy is able to drive out the mutant. Because the expected fitness of S_k increases in value as k gets larger, when k is large enough the expected fitness of S_k will exceed the expected fitness of S_1. In particular, for any proposed candidate $\bar{\epsilon} > 0$, it is the case that $W(S_k) > W(S_1)$ whenever $k > \frac{1-\bar{\epsilon}}{\bar{\epsilon}}$.

The final set-based notion of evolutionary stability Balkenborg and Schlag identify is the following:

Definition 10: Uniform Evolutionarily Stable Sets

Let L be a non-empty set of strategies for some game G. Then L is a *uniform evolutionarily stable set* if and only if

1. No strategy in L can spread, given any other strategy in L.
2. There exists a $\epsilon^* > 0$ such that, for any strategy σ in L and any strategy μ not in L, the strategy σ can ϵ^*-drive out the strategy μ.

Uniform ES sets are, of these three, the strongest conception of evolutionary stability. Here, there is a fixed threshold such that all members of the set L can drive out any mutant, as long as the mutant's frequency in the population lies below the threshold.

Notice that these notions of one 'strategy driving out another', or of a 'strategy spreading given the presence of the other', are not the same ideas as what was appealed to in Thomas's original definition of an ES set. In Definition 7, the mention of 'an ϵ-neighbourhood N around σ' was used to indicate how *close* the strategy μ was to σ, the intuition being that the mixed strategy $\frac{1}{8}S_1 + \frac{7}{8}S_2$ is closer to S_2 than to the mixed strategy $\frac{1}{2}S_1 + \frac{1}{2}S_2$.[22] As we've seen, Balkenborg and Schlag appeal instead to the idea of the average fitness of strategies in a population consisting of several types, where the ϵ refers to the *proportion* of the population following the mutant strategy. Sometimes these two different conceptual approaches don't make a material difference: one of the things that Balkenborg and Schlag prove is that, much of the time, their definition of *pointwise uniform evolutionarily stable sets* coincides with Thomas's definition of *evolutionarily stable sets*.

This raises a conceptual point which has been lurking in the background throughout this section: the difference between *mixed strategies* and *mixed*

---

[22] This conception of 'closeness' treats strategies as if they were nothing more than the *labels* on axes in Cartesian space. *Distance between strategies* reduces to normal spatial distance. What's the distance between $\sigma = \frac{1}{8}S_1 + \frac{7}{8}S_2$ and $\tau = S_2$? If σ is $\left(\frac{1}{8}, \frac{7}{8}\right)$ and τ is $(0, 1)$, then the distance is $\sqrt{\left(\frac{1}{8}-0\right)^2 + \left(\frac{7}{8}-1\right)^2} \approx 0.177$. But it isn't obvious that this is the right measure for the distance between strategies. Suppose we have a game where S_1 and S_2 yield payoffs of 1 whenever they are played against each other, and S_3 yields payoffs of 1 when played against itself and, if someone plays S_1 or S_2 against S_3, that player gets -100. In this case, does it seem right that the distance between the pure strategies S_1 and S_2 (i.e., 1) is the same as the distance between S_1 and S_3? If you think not, then you are rejecting the standard measure of strategic distance.

populations of strategies. It is important to keep these two notions distinct. The reason there can be conceptual slippage between the two is that many people, when first confronted with the idea of a mixed strategy, are a bit unsure about how to think about it. Earlier, when I first introduced the idea of mixed strategies on page 8, I suggested they could be thought of as a plan of action, where a person used a coin toss or a randomisation device to choose what to do while keeping their opponent uncertain. Most of the time, the first reaction of people who hear that suggestion is one of incredulity. On this point, Rubinstein (1991) remarked: 'the naïve interpretation of a mixed strategy as an action which is conditional on the outcome of a lottery executed by the player before the game, goes against our intuition.'[23] How, then, to make sense of a mixed strategy? An alternative proposal is what Rubinstein calls the *large population case*: 'One can think about a game as an interaction between large populations [...] In this context, a mixed strategy is viewed as the distribution of the pure choices in the population' (Rubinstein, 1991, p. 913).[24] In the large population case, playing against the mixed strategy $\frac{1}{2}S_1 + \frac{1}{2}S_2$ is to be interpreted as playing against a player selected at random from a population consisting of 50 per cent pure S_1-types and 50 per cent pure S_2-types.

Sometimes it makes no difference which interpretation of a mixed strategy one uses. In the two-ways-of-life game, from my personal point of view it makes no difference for my strategic deliberations if I am told that I will play the game against a single person using the mixed strategy $\frac{1}{2}S_1 + \frac{1}{2}S_2$ or against a single person selected at random from a population of 50 per cent people who always play S_1 and 50 per cent people who always play S_2. But can we always move so freely between the two interpretations?

Consider the coordination game of Figure 11. Clearly all three pure strategies are ESSes, because each of them is a strict Nash equilibrium. But now consider the following three mixed strategies:

$$\sigma_1 = \tfrac{1}{2}S_1 + \tfrac{1}{2}S_2$$
$$\sigma_2 = \tfrac{3}{8}S_1 + \tfrac{1}{2}S_2 + \tfrac{1}{8}S_3$$
$$\sigma_3 = \tfrac{1}{4}S_1 + \tfrac{9}{128}S_2 + \tfrac{87}{128}S_3.$$

Why those three mixed strategies? No particular reason other than they provide an illustration of some of the issues that arise when we move between the two interpretations of mixed strategies, as we'll now see.

---

[23] This is a measured claim. According to Oechssler (1997), an earlier working version of Rubinstein's paper stated that the naïve interpretation was 'ridiculous'.

[24] In his doctoral dissertation, Nash (1950a) proposed a similar interpretation of mixed strategies called the 'mass action' interpretation.

	S_1	S_2	S_3
S_1	(1,1)	(0,0)	(0,0)
S_2	(0,0)	(1,1)	(0,0)
S_3	(0,0)	(0,0)	(1,1)

Figure 11 A three-strategy coordination game.

Let's begin with the interpretation that treats mixed strategies as a self-contained plan of action involving a randomisation device. First, if the only possible strategies are σ_1, σ_2, and σ_3, then σ_1 is an ESS:

$$\pi(\sigma_1 \mid \sigma_1) = \tfrac{1}{2} > \tfrac{7}{16} = \pi(\sigma_2 \mid \sigma)$$
$$\pi(\sigma_1 \mid \sigma_1) = \tfrac{1}{2} > \tfrac{41}{256} = \pi(\sigma_3 \mid \sigma)$$

Second, in a *mixed population* consisting of *equal proportions* of σ_1, σ_2, and σ_3, it turns out that σ_1 has the greatest fitness of the three mixed strategies. Straightforward calculation shows, in such a population, that

$$W(\sigma_1) = \tfrac{1}{3}\pi(\sigma_1 \mid \sigma_1) + \tfrac{1}{3}\pi(\sigma_1 \mid \sigma_2) + \tfrac{1}{3}\pi(\sigma_1 \mid \sigma_3)$$
$$\approx 0.365$$

and that $W(\sigma_2) \approx 0.352$ and $W(\sigma_3) \approx 0.301$. That is, σ_1 is an ESS and is the unique best response (of the three possible strategies) given the composition of the population. In the absence of mutation, we would expect the number of σ_1-players to increase over time, as σ_1 has the highest expected fitness of all available strategies.

But now let's switch to the population interpretation of mixed strategies. In this case, we think of σ_1, σ_2, and σ_3 as representing self-contained sub-populations featuring the appropriate mix of pure strategies. Suppose, for simplicity, that the three sub-populations are of equal size. In this case, σ_1 would be a population containing equal numbers of players following S_1 and S_2. If I were to play a game against a player selected at random from that population, I would be indifferent between choosing S_1 and S_2: there is no unique best response. Similarly we reason that if I were pitted against a randomly selected player from the population corresponding to σ_2, the unique best response is S_2 because that's the pure strategy with the highest representation in the population. And, again, pitted against a player selected at random from the σ_3 population, the unique best response is S_3.

Now suppose the three sub-populations corresponding to σ_1, σ_3, and σ_3 merge, as before. As the sub-populations are of equal size, the resulting aggregate population would have the three pure strategies represented in proportions according to the weighted sum

$$\tfrac{1}{3}\sigma_1 + \tfrac{1}{3}\sigma_2 + \tfrac{1}{3}\sigma_3 = \left(\tfrac{3}{8}, \tfrac{137}{384}, \tfrac{103}{384}\right)$$

$$\approx (0.375,\ 0.357,\ 0.268).$$

Notice what has happened: even though S_1 is not the unique best reply to *any* of the sub-populations corresponding to σ_1, σ_2, or σ_3, it is the unique best reply to the merged population! In the absence of mutation, we would expect the number of S_1-players to increase over time, as S_1 has the highest expected fitness of all available strategies.

What this example shows is that sometimes it *can* make a difference whether we interpret a mixed strategy as a plan of action or as a mixed population of pure strategies. According to the preceding discussion, whether we would expect S_1 or σ_1 to drive out other strategies present depends on which interpretation of mixed strategies we use. And I think there is another lesson we can draw. I suggest this example shows the limits of thinking, in the abstract, about evolutionary stability based on pairwise comparisons of payoffs and back-of-the-envelope calculations of expected fitness. A proper analysis of evolutionary stability requires that we get into the details of evolutionary dynamics: population structure, individual interactions, and selection mechanisms. That is the topic of the next section.

3 Continuous Dynamical Models of Evolutionary Games

3.1 Introduction

One of the distinguishing features of the analysis of evolutionary games in Section 2 is that all of the concepts employed – a Nash equilibrium, an evolutionarily stable strategy, or an evolutionarily stable set – were *static*. The concepts only checked whether or not a certain set of conditions, defined in terms of the game's payoff matrix, held (e.g., Definition 1). Even Balkenborg and Schlag's characterisations of evolutionary stable sets, which mentioned one strategy 'driving out another', were defined in terms of expected payoffs. There was no detailed attempt to see *how* a population of players might reach a state where everyone followed an evolutionarily stable strategy, or followed one strategy from an evolutionarily stable set. This is an important question because evolution is a selection process subject to random shocks, both from within the population (e.g., who happens to interact with who) and from outside the population (e.g., whether a global event occurs which generally affects the fitness of individual strategies). Given this, it would be useful to be able to model what happens when the population is not already at an equilibrium state. As we will see, much recent work in evolutionary game theory gives us the tools necessary to construct dynamical models of evolving populations.

A further benefit is that these dynamical models allow for multiple interpretations: the underlying mathematics can be viewed as models of either biological or cultural evolution.

These two approaches to evolutionary game theory – the static and the dynamic – were present from the very beginning of the field. Around the time that Maynard Smith and Price were developing their concept of an evolutionarily stable strategy, the mathematical biologists Taylor and Jonker (1978) were modelling an evolving population as a *dynamical system*. In mathematics, a dynamical system provides a description of the time-dependent state of a system, where 'the state of the system' is a complete specification of all properties required to uniquely characterise the system. We can represent the *current state* of the system as a point in a multidimensional state space, each point representing one possible configuration. The *future state* of the system is given by a function applied to the current state. Standard examples of dynamical systems in classical physics are the pendulum, the trajectory of projectiles near the Earth's surface, and the motion of bodies under mutual gravitational attraction. In these cases, the state of each system is given by a vector specifying the mass of each object, the location of the centre of mass of each object, and the velocity of each object. The future state of each system is determined by Newton's laws of motion and the law of gravity.

Evolving populations can be conceived as a dynamical system. The simplest models assume that the only thing that matters is the relative proportion of each trait in the population, effectively treating all individuals as fungible. In these models, the state of the evolutionary system is a point $\vec{p}(t) = \langle p_1(t), \dots, p_n(t) \rangle$ in $\mathbb{R}^n$, where each $p_i(t)$ is a number between 0 and 1 indicating the proportion of the population following strategy S_i at time t.[25] The dynamical laws of the evolutionary system state how the strategy frequencies $\vec{p}$ change over time, tracing out a path in state space.[26] Given a particular initial condition $\vec{p}(0)$, different evolutionary models may predict different future trajectories, leading to substantially different long-term outcomes starting from the same state. It is, of course, also possible that different evolutionary models may predict substantially the *same* long-term outcome, starting from the same state, but following different paths to get there.

---

[25] In what follows, I will sometimes drop the explicit time-dependency (e.g., writing p_i instead of $p_i(t)$) when the time-dependency is clear from the context. This helps reduce notational clutter.

[26] The path is not free to travel throughout all of $\mathbb{R}^n$. Since $\vec{p}$ is subject to the constraint that $p_i \geq 0$, for all i, and that $\sum_{i=1}^{k} p_i = 1$, the path traced lives on a hyperplane of $\mathbb{R}^n$.

Although evolutionary models which only pay attention to strategy frequencies are highly idealised and ignore many aspects which could matter,[27] they are a useful place to begin exploring the complexities and nuances which arise when we try to see how the static concepts of evolutionary stability map onto dynamic concepts of evolutionary stability. What we will find is that even these idealised models yield surprising results which align imperfectly with the static analysis from Section 2, (See Mohseni, 2015 for a very nice discussion of the details.) The first such idealised model we will consider is called the *replicator dynamics*.

3.2 The Replicator Dynamics

Let $S = \{S_1, \ldots, S_k\}$ be the set of pure strategies for the game under consideration, where we denote the payoff function by π, as before. Let $0 \leq p_i(t) \leq 1$ denote the proportion of the population which follows strategy S_i at time t. Assuming that the only thing which matters, from an evolutionary perspective, is how many individuals follow a given strategy, then the complete state of the system at time t is represented by the vector $\vec{p}(t) = \langle p_1(t), \ldots, p_k(t) \rangle$, where $\sum_{i=1}^{k} p_i(t) = 1$. In addition, assume that individuals interact *at random* where the probability of interacting with someone following S_i is simply p_i. This is a reasonable assumption if the population is very large, types are evenly distributed throughout, and pair formation occurs by sampling.

Under these assumptions, it is straightforward to calculate the expected fitness of S_i given the state of the population at time t, a quantity typically denoted $W_i(t)$:

$$W_i(t) = \sum_{j=1}^{k} p_j(t) \cdot \pi\left(S_i \mid S_j\right).$$

Why does it take on this value? A player following S_i has p_j chance of interacting with someone following S_j, which would yield a payoff of $\pi(S_i \mid S_j)$. The total expected fitness, then, is the indicated sum. Once we know the expected fitness of each individual strategy in the population at a given time, it is equally straightforward to calculate the average fitness of the population at that time:

$$\overline{W}(t) = \sum_{i=1}^{k} p_i(t) \cdot W_i(t).$$

Why do we care about the average fitness of the population? The principle of natural selection states that individuals with higher reproductive fitness are

---

[27] Such as the spatial location of individuals, group structure, the particular learning history of individuals, stochastic effects of the environment, and so on.

more likely to have their traits represented in subsequent generations. As a first approximation, the average fitness of the population provides a baseline against which the fitness of individual strategies can be compared, determining whether the number of individuals following any specific strategy will grow or shrink in number.

Now we are in a position to define the replicator dynamics. Since the state of the system is fully specified by $\vec{p}$, the dynamics only need to state how each of the p_i change over time. The functional form of the replicator dynamics is as follows:

Definition 11: The Replicator Dynamics

$$\frac{dp_i}{dt} = p_i \cdot \left(W_i - \overline{W} \right), \text{ for } i = 1, \ldots, k.$$

We'll see an explicit derivation of the replicator dynamics shortly, but let's first take a moment to reflect on what that equation says.[28] If the expected fitness of S_i is greater than the average fitness of the population, the right-hand side of the replicator equation is positive, which means that the rate of change of p_i is positive. A positive rate of change means p_i will increase. This makes sense because if the expected fitness of S_i is greater than the average fitness of the population, it must be the case that S_i is more fit than at least one other strategy (not necessarily *all* others) in the population and will grow at the expense of the less fit strategies. If the expected fitness of S_i is less than the average fitness of the population, the rate of change of p_i is negative and hence will shrink over time.[29] In the case where the expected fitness of S_i equals the average fitness of the population, the right-hand side equals 0 and there is no change in p_i at all.

---

[28] Strictly speaking, it is a system of equations because $i = 1, \ldots, n$.

[29] Keep in mind that p_i represents the *proportion* of the population following S_i. If there is plenty of food in the environment and no predators and no limits on the carrying capacity of the environment, a greater-than-average fitness means that the strategy's population proportion will increase even if there is no explicit selection (i.e., death) of offspring of *any* strategy. All strategies could increase in number, in absolute terms, while some strategies see a decrease in their proportion of the population. Suppose we have 1,000 S_1-types and 20 S_2-types in a population. Then $p_1 \approx 0.980$ and $p_2 \approx 0.020$. If S_1 has an expected fitness of 1.1 and S_2 has an expected fitness of 1.2, then the next generation will have 1,100 S_1-types and 24 S_2-types, with $p_1 \approx 0.979$ and $p_2 \approx 0.021$. Somewhat more counterintuitive is the fact that, if the environment has infinite carrying capacity (which means there is no limit on the number of individuals which can survive), then the proportion of S_1-types, $p_1(t) = \frac{1000 \cdot (1.1)^t}{1000 \cdot (1.2)^t + 20 \cdot (1.1)^t}$ still converges to 0 as $t \to \infty$ even though the absolute number of S_1-types increases without bound!

The other thing to note about the replicator dynamics is that p_i appears on the right-hand side of the equation. This means that the rate of growth of p_i is proportional to itself. That makes sense, because the more S_i-individuals there are in a population, the more offspring they will have at a given fitness level. But there are two other important implications as well. The first is that if $p_i(t') > 0$, for any particular time t', then $p_i(t) > 0$ for all times t. That's because the replicator dynamics only says what the *rate of change* is: even if the rate of change of p_i is always negative, p_i will get arbitrarily close to 0 but will never actually equal 0.[30] The second implication is that if $p_i(t') = 0$, for any particular time t', then $p_i(t) = 0$ for all times t. One limitation of the replicator dynamics is it cannot introduce *new* strategies into the population.

3.2.1 A Model of Biological Evolution

Now that we talked through the interpretation of the replicator dynamics, let us see how to derive it as a model of biological evolution. Denote the total number of individuals in the population following S_i by the integer $n_i \geq 0$. The total number of individuals in the entire population is thus $N = n_1 + \cdots + n_k$, and the proportion of strategy S_i is simply $p_i = \frac{n_i}{N}$.[31]

Assume that reproduction takes places continuously via cloning. That means there aren't well-defined breaks between generations, and that all children follow the same strategy as their parent. Assume also that the average birth rate for individuals following S_i is $F_0 + W_i$, where F_0 denotes a basic background fitness shared by all individuals. (Think of this as a common fecundity all have prior to any interaction.) If, in addition, individuals have a common death rate of $\delta > 0$, then the rate of change in the number of individuals following S_i is

$$\frac{dn_i}{dt} = (F_0 + W_i - \delta) n_i.$$

---

[30] Think of the function $f(x) = \frac{1}{x}$. No matter how large x gets, $f(x)$ will always remain positive, even though it will get very small as x becomes very large.

[31] There is a subtle mathematical modelling point which needs to be mentioned. All of the n_i, and by implication N, need to be large, positive integers. That is because we are representing the population using a vector of k real numbers, with each p_i between 0 and 1, and using a system of differential equations to describe the rate of change in the p_i. This means the p_i vary as continuous quantities. In order for small changes in p_i to be meaningful, those small changes must reflect actual changes in the proportion of the population following S_i, and this curiously requires the population to be large. For example, suppose p_1 changed from 0.1274286 to 0.127429 over a short interval of time. That could correspond to n_1 changing from 1,274,286 individuals following S_1 to 1,274,290 individuals, in a total population of size 100,000,000. (Alternatively, it could also result from S_1 moving from 12,742,860 out of 100,000,000 individuals to 127,429 out of 10,000,000 individuals – perhaps due to some catastrophic change.) The counter-intuitive point is that the *smaller* the change in the value of p_i, the *larger* the population must be in order for that to be meaningful. In the limit, we have an infinite population and need to shift from talk of 'the number of individuals' following a strategy to talk of population measures.

All of these assumptions can be challenged, of course. Not all species lack well-defined generational breaks, for example. And there are a number of reasons for challenging the claim that all individuals, regardless of their type, have a common fecundity F_0 and that the death rate δ is common. Yet these assumptions do introduce a greater level of generality to the derivation because the resulting set of equations we obtain, somewhat surprisingly, turns out to not depend on the specific value of either F_0 or δ.

Using the fact that $p_i N = n_i$, a bit of calculus yields

$$\frac{dp_i}{dt} N + p_i \frac{dN}{dt} = \frac{dn_i}{dt}$$

where

$$\frac{dN}{dt} = \frac{d}{dt}\left(\sum_{j=1}^{k} n_j\right) = \sum_{j=1}^{k} \frac{dn_j}{dt} = \sum_{j=1}^{k} (F_0 + W_j - \delta) n_j$$

$$= (F_0 - \delta) \sum_{j=1}^{k} n_j + \sum_{j=1}^{k} W_j n_j$$

$$= (F_0 - \delta)N + \sum_{j=1}^{k} W_j p_j N$$

$$= \left(F_0 - \delta + \bar{W}\right) N.$$

Given this,

$$p_i \frac{dN}{dt} = \left(F_0 - \delta + \bar{W}\right) n_i.$$

And so

$$\frac{dp_i}{dt} = \frac{1}{N}\left(\frac{dn_i}{dt} - p_i \frac{dN}{dt}\right)$$

$$= \frac{1}{N}\left((F_0 + W_i - \delta)n_i - \left[(F_0 - \delta + \bar{W})n_i\right]\right)$$

$$= p_i \left(W_i - \bar{W}\right).$$

This shows how the mathematics of the replicator dynamics can be interpreted as providing a model of biological evolution. The pattern of individual births and deaths across all individuals, when aggregated, yields the replicator dynamics as population-level laws describing how strategy frequencies change over time.

3.2.2 A Model of Cultural Evolution

What would be the cultural evolutionary analogue for the process of births and deaths we just used? In the cultural evolutionary case, think of the population of individuals as constant and fixed over time, with change occurring amongst the *beliefs* and *behaviours* of those individuals. The idea is that beliefs, and the behaviours resulting from those beliefs, are the items being selected for or against. An agent follows a particular strategy in a game-theoretic context and, based on its performance, either will continue to use it or will switch to a different strategy. A rational agent will adjust their beliefs in light of the evidence received, according to some to-be-specified process of *belief revision*. The belief revision process is the analogue of natural selection, and the individual beliefs (i.e., strategies) are the units of selection.

As the cultural evolutionary case is a little more complicated than the biological case, I won't provide details here. (But interested readers should consult both Björnerstedt and Weibull, 1999 and Alexander, 2007 for more information.) Instead, we will discuss one general approach, due to Sandholm (2010), that allows us to derive dynamics for a number of cultural evolutionary models. One of the benefits of Sandholm's method, for our purposes, is that it allows us to describe the relationship between individual behaviour and the resulting population dynamics without getting bogged down in the mathematical details.

In his groundbreaking work *Population Games and Evolutionary Dynamics*,[32] Sandholm presents a general framework for modelling the foundations of evolutionary games, showing how many different population models can be derived by aggregating individual choice behaviour. More precisely, we are interested in identifying how changes in the frequency of strategies, at the level of the *population*, depend on the specific belief revision process used by *individuals* to change the particular strategy they will use in future rounds of play.[33] Different revision processes can be shown to yield, at the population level, different dynamical laws describing how strategy frequencies change over time. Some revision processes yield the replicator dynamics, and others do not.

Let us introduce some notation for greater precision. As before, we use $\vec{p} = \langle p_1, \ldots, p_k \rangle$ to denote the current population state. Let $\vec{W} = \langle W_1, \ldots, W_k \rangle$ be the vector of expected payoffs of each strategy, given the current state of the population. One important subtlety is that $\vec{W}$ may have nonzero values for W_i even if $p_i = 0$. Why? Because the value of W_i depends only on the payoff matrix

---

[32] I mention the title for two reasons: first, it is a book of exceptional richness and depth and highly recommended (but it demands a mathematically sophisticated reader); second, the title is a deliberate nod to an earlier work by Hofbauer and Sigmund (2002) named *Evolutionary Games and Population Dynamics* – and so the two books are easily confused unless one is aware of the similarity.

[33] Sandholm calls this a *revision protocol*, but I will stick to the terminology used within decision theory.

for the game and the current population state; that is, W_i should be read as the answer to the question, 'What payoff could a player expect to receive *if* they used S_i in the state $\vec{p}$?' This matters because some belief revision processes use this information to introduce strategies into the population even if they are initially absent, in contrast to the replicator dynamics.

According to Sandholm, the information contained within $\vec{W}$ and $\vec{p}$, along with the belief revision process, determines the rate at which individuals switch from one strategy to another. In what follows, I will use ρ to denote the belief revision process under discussion and ρ_{ij}, where $i \neq j$, to denote the specific rate at which an individual who uses ρ switches from S_i to S_j. The value ρ_{ii} is thus the rate at which people continue to use S_i.

Suppose that some individual A is deliberating on whether to revise their strategy, and they do so as follows. First, A samples another player, call them B, at random from the population. If B's payoff in the last round of play exceeded A's payoff, then A will adopt B's strategy with a probability proportional to the *difference* in payoffs. This belief revision process is known as *imitate with probability proportional to success*. When the population is very large, the chance of sampling someone following S_j is simply p_j, and the payoff received by the person sampled will, on average, be W_j. Thus we have

$$\rho_{ij} = p_j \cdot \left[W_j - W_i \right]_+$$

where $[x]_+ = \max(x, 0)$. It can be shown (see Sandholm, 2010) that a population of individuals who revise their strategy in this fashion will, at the population level, generate changes to the overall distribution of strategies exactly as described by the replicator dynamics.

Here we see one of the great benefits of evolutionary game theory for the social sciences. Whereas *traditional* game theory was concerned with the strategic behaviour of perfectly rational individuals, all of whom have complete knowledge of the game and can solve complex maximisation problems, *evolutionary* game theory can be viewed as being concerned with the strategic behaviour of *boundedly* rational individuals. A boundedly rational individual is someone who is trying to make the best decision possible, but who falls short of the ideal rational agent so frequently assumed in game theory and economics. There are a number of reasons why someone might be boundedly rational: limits on the amount of information available, or on one's ability to compute a solution, or limits on the amount of time available for deliberation, all can make a difference.[34] When might it be rational for a person to revise their strategic

---

[34] It's important to appreciate that the methods of evolutionary game theory are not the only way to address these concerns. There is a large literature, developed within the traditional theory of games, which examines both such questions. Those interested in the connection

behaviour using imitate with probability proportional to success? One instance would be when a player doesn't actually know the form of the payoff matrix. Isn't it implausible to think that a player wouldn't know the form of the payoff matrix? Not at all. In most of real life, we don't actually know the payoff matrix for the 'game' we are playing. In such a case, it can be rational for a player with limited information to periodically sample someone from the population, compare their respective performances, and then imitate when it seems potentially profitable. We will revisit imitative dynamics in Section 4.

3.2.3 Some Simple Games

Figure 12 illustrates,[35] for four different games, the evolutionary behaviour of a population under the replicator dynamics. The payoff matrix for each of the four games appears on the left, and a diagram showing the evolutionary flows on the right. The diagrams are interpreted as follows: the line segment represents the interval $[0, 1]$, where a point $q \in [0, 1]$ on the line indicates the proportion of the population following strategy S_2. Under such a representation, the leftmost point on the line segment, corresponding to $q = 0$, represents the state where the entire population follows strategy S_1. The rightmost point on the line segment, where $q = 1$, is the statement where the entire population follows S_2, and points in the middle represent a mixed state of the population where both strategies are present.

Circles in the diagram represent *fixed points* of the dynamical system. A fixed point is, as the name suggests, a state where the system is in equilibrium,[36] and none of the variables change in value. A fixed point, for the systems studied here, represents a population in evolutionary stasis; an evolutionary

---

between bounded rationality and game theory should consider Samuelson (1996), which provides a short introduction to how bounded rationality can be approached in both cooperative and noncooperative game theory. (In section 5 of that paper, Samuelson also discusses evolutionary game theory.) Another excellent resource is Rubinstein (1998), although the connection between bounded rationality and game theory is only a subset of his concerns. A classic source regarding games of incomplete information is the three-paper set by Harsanyi (1967a,b,c), but they are mathematically challenging. The general study of games where the assumptions of who knows what are made explicit is the subfield known as *epistemic game theory*, to which Pacuit and Roy (2017) provide an excellent introduction.

[35] Many of the figures in this section were produced using *Dynamo: Diagrams for Evolutionary Game Dynamics*, an open-source package for *Mathematica*. Many thanks to Francisco Franchetti and William H. Sandholm for making this software freely available. See Franchetti and Sandholm (2013) for further details.

[36] The term 'equilibrium', as used here, should not be confused with that of a Nash equilibrium. A Nash equilibrium refers to an individual strategy which is a best response to the strategies used by all other players. An equilibrium point of the replicator dynamics (or other dynamics) is a point where $\frac{dp_i}{dt} = 0$, for all i. There is an important question as to what, if any, relationship exists between Nash equilibrium strategies and equilibrium points of a population, which we will consider in what follows.

'stalemate', so to speak, where the relative fitness of all strategies is the same. A black circle represents a fixed point which is stable (a concept which we will make more precise later). A white circle represents an unstable fixed point, where any deviation whatsoever from the respective population proportions will result in the system not returning to that state.[37]

In discussing the derivation of the replicator dynamics, we have often talked about what happens at the individual level using the language of chance. Under the biological interpretation, we spoke of the birth and death rate of individuals. Under the cultural interpretation, we spoke of the chance that A samples someone from the population who uses strategy S_j, and of the chance that A adopts B's strategy. At the population level, all of this indeterminacy washes out, leaving a purely *deterministic* system of dynamical laws.

One important property of deterministic dynamical systems is known as the 'No-Intersection Theorem' (Hilborn, 2000). According to this theorem, two different solutions to the replicator dynamics (and other dynamics we will consider in this section) cannot intersect, although they may converge to the same point in the limit as $t \rightarrow \infty$. (This latter case is what happens in the Hawk–Dove game of Figure 12c; there, the paths which approach the stable fixed point featuring $1-\frac{V}{C}$ Doves don't ever reach that point in finite time, although they can get arbitrarily close.) In addition, no solution may intersect itself at a finite number of points.[38]

Figure 12a shows the evolutionary outcome, under the replicator dynamics, for the Prisoner's Dilemma. This well-known game, first discussed by Flood (1952), represents the strategic problem of cooperation and is typically motivated by the following vignette. Suppose that two suspected criminals are caught by the authorities and are put into separate holding cells, unable to communicate with each other.[39] The investigator approaches one of the suspects and tells them, "Look – we know you are guilty, but the evidence we have is a bit

---

[37] This might strike one as odd: isn't evolution a process driven by chance? Doesn't the revision protocol underlying the replicator dynamics refer to random sampling of individuals from the population? The answer to both questions is *yes*, and thus we are confronted with one of the curious consequences of the modelling assumptions underlying the replicator dynamics – and some other population dynamics we will consider as well. Although chance events feature at the individual level, as the population increases in size, the behaviour of the population *in aggregate* converges to a deterministic process.

[38] The 'finite number of points' qualification is needed because, as we will see, it is possible for solutions to the replicator dynamics – and other dynamics – to take the form of closed loops, or cycles. In these cases, a solution will self-intersect at an infinite number of points.

[39] Strictly speaking, it's not necessary for the suspected criminals to be put into separate cells with no possibility of communication: in the absence of any enforceable commitment mechanism, any talk between the two suspects would be 'cheap talk', so called because it cannot affect the payoffs of the game. In the absence of an enforceable commitment mechanism, even if I *say* that I *promise* not to turn state's evidence, that promise is meaningless because there's no way to hold me accountable to what I say if I do something different. And if one suspect is

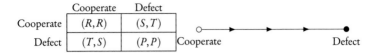

(a) The Prisoner's Dilemma, where $T > R > P > S$ and $\frac{T+S}{2} < R$.

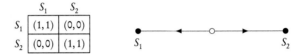

(b) A pure coordination game.

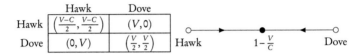

(c) The Hawk–Dove game, where $0 < V < C$.

	Do this	Do that
Do this	(0,0)	(0,0)
Do that	(0,0)	(0,0)

(d) The game of life.

Figure 12 The four possible evolutionary trajectory patterns for two-strategy games under the replicator dynamics. Modifications to the payoffs for games (b) and (c) may change where the unstable/stable fixed point in the middle is located.

thin. If you agree to turn state's evidence, we will drop all charges against you and use your testimony to convict your partner; however, you should know that we're making the same offer to your partner. Think fast about what you want to do, as this deal won't be around for long."

The payoff matrix of Figure 12a shows the general payoff matrix for this type of problem. The strategy labels are from the point of view of the joint enterprise: 'Cooperate' means that the person stays silent and does not turn state's

---

worried that, after turning state's evidence, the other suspect will hunt them down after being released and take revenge, that should already be reflected in the payoffs of the game. (If one suspect *is* concerned about that prospect, then the game will likely not have the structure of the prisoner's dilemma!) In this vignette, I adopt a common approach to describing the prisoner's dilemma even though it includes unnecessary restrictions. That is simply because, for people encountering the game for the first time, the strategic intuitions elicited regarding this problem are not encumbered by complications (like talking!) which might *appear* to matter, even though they are actually irrelevant.

evidence, and 'Defect' means that the person rats out their partner. The best possible payoff, from each person's point of view, occurs when they Defect against a Cooperator (the defector gets off scot-free with no criminal record or jail time). The second-best payoff occurs when both people Cooperate, as staying silent means that the investigator is left only with the thin evidence against them (some chance of being convicted, but not very high, and both people are in the same boat). The third-best payoff occurs when both people Defect, on the grounds that the prosecutor chooses to offer shorter prison terms when both turn state's evidence. The worst payoff is when a person chooses to Cooperate, staying silent while their partner Defects. The payoff labels are chosen to reflect this structure: the (T)emptation of getting off scot-free at your partner's expense is greater than the (R)eward both receive by staying silent, which in turn is greater than the (P)unishment both receive when they turn state's evidence against each other, which is greater than the (S)ucker's payoff, staying silent when your partner turns state's evidence. The second requirement, that $\frac{T+S}{2} < R$, is a technical point to handle cases where the Prisoner's Dilemma is *repeated* many times: a true problem of cooperation should have always cooperating be better than taking turns defecting on each other.

In the Prisoner's Dilemma, the strategy Defect strictly dominates Cooperate. Hence there is exactly one Nash equilibrium in which both players choose Defect. The diagram on the right of Figure 12a shows that, under the replicator dynamics, the long-term evolutionary outcome agrees with the traditional game theoretic analysis: any population state containing some Cooperators will lead to a future population state containing a lower proportion of Cooperators, converging to the state of All-Defect in the limit. The population state containing only Cooperators is a fixed, but unstable, point under the replicator dynamics. (Recall the earlier remark about the replicator dynamics not being able to introduce strategies into the population if they are initially absent.) The All-Cooperate state is unstable because, if we imagine making an intervention and inserting any $\epsilon > 0$ of Defectors into the population, those Defectors will grow in proportion until they dominate the whole population.

The remaining three games of Figure 12 illustrate the other evolutionary possibilities for two-player, two-strategy games. Figure 12b shows a coordination game, and we see that replicator dynamics yields an outcome that is reasonable: in a population with more of one type than the other, the type with the initial higher proportion will drive the other one to extinction. If the population is perfectly balanced with both types equally represented, then evolution has no reason for selecting one type over the other, as both have equal fitness; but this fixed point is unstable. Figure 12c shows the Hawk–Dove game, where under the replicator dynamics the population will converge to a stable point containing a mix of both Hawks and Doves, with $\frac{V}{C}$ of the population following the strategy Hawk.

Finally, Figure 12d shows a degenerate game discussed by Skyrms (2010). The game can be interpreted as representing the situation of existentialist angst, since none of the available options make a material difference to the outcome.[40] Here, every population state is a fixed point: because all strategies have the same payoff, there is no selection pressure at all and the population will remain in whatever state it is in. However, there is a subtle difference between the type of fixed point in this game and the fixed point found in the Prisoner's Dilemma and the Hawk–Dove game. In those games, any (reasonable) displacement from the fixed point will result in the system eventually converging back to the fixed point.[41] In the game of life, any displacement will not result in the system returning to its prior state; but, then again, there will be no further movement *away*.

3.2.4 Stability Concepts

What the game of life illustrates is that there are two different notions of dynamic stability which need to be distinguished. The type of stability present in the game of life is known as *Lyapunov stability*, named after the Russian mathematician Aleksandr Lyapunov. A system is said to be Lyapunov stable at a point $\vec{p}$ if, for any initial point $\vec{q}_0$ sufficiently close to $\vec{p}$, the system will remain close to $\vec{p}$ at all future times. Intuitively, there is no local 'push' away from the fixed point. It is in this sense that the game of life is stable – obviously trivially so, given the nature of the payoffs, but we will see less trivial cases in a moment.

The second type of stability is found in both the Prisoner's Dilemma and the Hawk–Dove game. In these two games, it's not only true that any point 'close' to the fixed point – All-Defect, in the Prisoner's Dilemma, and the mixed population state in the Hawk–Dove game – will remain close, in the future. Here, any point close to the fixed point will converge *back* to it. Intuitively, we might say there is a local 'pull' back towards the fixed point. This type of stability is known as *asymptotic stability*. The set of all points which converge to a fixed point $\vec{p}$ are known as the *basin of attraction* of $\vec{p}$. For both the Prisoner's Dilemma and the Hawk–Dove game, the basin of attraction for the asymptotically stable point is the entire set of all population states in which both strategies are represented.

---

[40] The name 'the game of life' is thus intended as a wry joke. It should not be confused with John Horton Conway's Game of Life – a specific two-dimensional cellular automaton capable of universal computation – which is a very different thing.

[41] The qualification about any 'reasonable' displacement is required because an extremal displacement that moved the population to the All-Cooperate state would remain there.

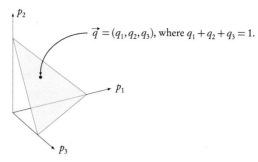

Figure 13 The simplex for the three-strategy replicator dynamics.

Let us now turn to consider games with three strategies, and see how these concepts of dynamical stability play out there. In games with three (or more) strategies, there are many more possibilities regarding the evolutionary outcomes. (See Bomze, 1983, for a detailed catalogue of all possible cases for the three-strategy replicator dynamics.) But, before getting into some of the details, we first need to discuss a standard representation used in the evolutionary game theoretic literature.

For a game with three strategies, the population state is specified by a vector $\vec{p} = (p_1, p_2, p_3)$, where $p_1 + p_2 + p_3 = 1$ and $p_i \geq 0$ for all i. Figure 13 shows the set of all points which satisfy this restriction as a shaded gray region in three-dimensional space. It will help to recall that $x + y + z = 1$ is an equation of a plane in three-dimensional space, and the requirement that all of the three quantities are positive means that we are only interested in the portion of the plane shown. Each population state corresponds to a *single* point in the gray triangular region, known as the *simplex*.[42] From an initial condition, like that of $\vec{q}$ shown, as the population evolves under the replicator dynamics – or, for that matter, any other dynamics – it will trace out a *path* in the simplex.[43] Because the three-dimensional orientation of the simplex is unnecessary information, normally the simplex is drawn so that we face it directly, as an equilateral triangle.

Figure 14 shows several evolutionary trajectories for the game of Rock-Paper-Scissors under the replicator dynamics. Each vertex of the simplex, corresponding to a population consisting entirely of individuals playing Rock (or Paper, or Scissors), is an unstable fixed point. Introducing any $\epsilon > 0$ of players following Paper into an All-Rock population will cause Rock to be

---

[42] This term is from geometry, where a 'simplex' is the generalisation of the concept of a triangle to higher dimensions.

[43] The path is connected because solutions to the replicator dynamics are guaranteed to be continuous, as they have to be differentiable.

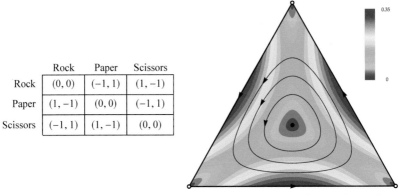

	Rock	Paper	Scissors
Rock	$(0, 0)$	$(-1, 1)$	$(1, -1)$
Paper	$(1, -1)$	$(0, 0)$	$(-1, 1)$
Scissors	$(-1, 1)$	$(1, -1)$	$(0, 0)$

Figure 14 The game of Rock-Paper-Scissors under the replicator dynamics. The shading represents the speed of motion of the population, as indicated in the legend.

driven to extinction. (And a similar point holds for the All-Paper and All-Scissors states, provided the suitable corresponding strategy is introduced.) The arrows on the edges of the simplex indicate these evolutionary trajectories.

In the interior of the simplex, where all three strategies are present, we see that almost all of the trajectories are cycles. (The one exception is the population state where all strategies are present in equal proportion.) The cyclical behaviour arises from the structure of the game, since there is no strategy which is always the best to play. What strategy counts as the 'best' one is entirely context-dependent. In a population consisting mostly of individuals playing Rock, it is best to play Paper. In such a population, Paper will have the highest expected fitness and so increase over time. However, at some point there will be so many individuals playing Paper that the strategy Scissors will become the best strategy to play, and so Scissors will start to increase. And, of course, at some point, there will be so many individuals playing Scissors that Rock will once again become the best strategy to play. The fact that the population will return to *exactly* the same state it was at previously is due to the precisely balanced payoffs: the fitness gain from winning is exactly equal to the fitness loss from losing.

The point $\vec{p}_e = \left\langle \frac{1}{3}, \frac{1}{3}, \frac{1}{3} \right\rangle$ is a Lyapunov stable fixed point. Small displacements away from $\vec{p}_e$ will continue to orbit around $\vec{p}_e$, never moving further away than a certain maximal distance, depending on the orbit.[44] This is a more interesting example of Lyapunov stability than we first saw in the game of life.

---

[44] Inspection of the diagram will show that the orbits aren't perfect circles around $\vec{p}_e$, but instead move closer and further away from $\vec{p}_e$ depending on where in the orbit we are. The point of maximal distance on an orbit from $\vec{p}_e$ corresponds to those states with the highest proportion of either Rock, or Paper, or Scissors.

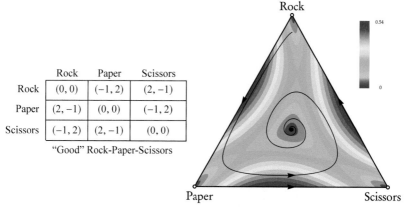

	Rock	Paper	Scissors
Rock	$(0,0)$	$(-1,2)$	$(2,-1)$
Paper	$(2,-1)$	$(0,0)$	$(-1,2)$
Scissors	$(-1,2)$	$(2,-1)$	$(0,0)$

"Good" Rock-Paper-Scissors

(a) One variant of Rock-Paper-Scissors with the interior fixed point asymptotically stable.

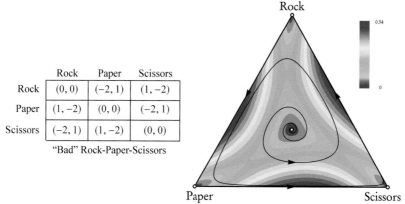

	Rock	Paper	Scissors
Rock	$(0,0)$	$(-2,1)$	$(1,-2)$
Paper	$(1,-2)$	$(0,0)$	$(-2,1)$
Scissors	$(-2,1)$	$(1,-2)$	$(0,0)$

"Bad" Rock-Paper-Scissors

(b) Another variant of Rock-Paper-Scissors with an unstable interior fixed point.

Figure 15 Two other versions of Rock-Paper-Scissors under the replicator dynamics.

But, as Figure 15 shows, these particular evolutionary outcomes are highly dependent on the payoff matrix for Rock-Paper-Scissors having the win/lose payoffs perfectly symmetric. The moment we disrupt the perfect symmetry, the evolutionary behaviour changes radically. In Figure 15a, the state $\vec{p}_e$ is asymptotically stable, with the entire interior of the simplex as its basin of attraction. In Figure 15b, the state $\vec{p}_e$ is no longer Lyapunov stable, although it is still a fixed point of the replicator dynamics. Here, evolution moves the population away from $\vec{p}_e$, spiralling ever closer to the boundary of the simplex, without converging to anything even in the limit.

At this point, we can begin to appreciate the nuanced relationship that exists between the Nash equilibrium solution concept, the concept of an evolutionary stable strategy, and the various concepts of dynamic stability. In all the games considered so far – the Prisoner's Dilemma, the pure coordination game, the Hawk–Dove game, the game of life, and Rock-Paper-Scissors – all of the Nash equilibria of the game correspond to fixed points of the replicator dynamics, if we interpret the Nash equilibrium *strategy* as a distribution over pure strategies in the *population*. In the Prisoner's Dilemma, the Nash equilibrium strategy of Defect corresponds to the All-Defect state. In the Hawk–Dove game, when $0 < V < C$, the Nash equilibrium in mixed-strategies is $\sigma^* = \frac{V}{C}$ Hawk $+ \left(1 - \frac{V}{C}\right)$ Dove. The corresponding population state, with $\frac{V}{C}$ Hawks and $1 - \frac{V}{C}$ Doves, is a fixed point of the replicator dynamics. In the pure coordination game, all three Nash equilibria (two in pure strategies and one in mixed-strategies) are fixed points of the replicator dynamics. In Rock-Paper-Scissors, the same results are found to hold, too.

In addition, we see that those fixed points which correspond to ESS are asymptotically stable with significant basins of attraction. All-Defect, in the Prisoner's Dilemma, is an attractor for the entire interior, as is the mixed population state of the Hawk–Dove game. In the pure coordination game, each of the two ESSes attract half of the total state space, with the unstable fixed point corresponding to the Nash equilibrium (which isn't an ESS) dividing the two regions.

But the fit between fixed points and Nash equilibria is imperfect. There are some fixed points of the replicator dynamics which are neither ESS nor a Nash equilibrium of the underlying game. The All-Cooperate state of the Prisoner's Dilemma is an example of this, as is All-Hawk and All-Dove of the Hawk–Dove game. What of the game of life, with its infinitely many fixed points? As we've noted, all of those fixed points are Lyapunov stable. Is it a coincidence that every strategy, both pure and mixed, is also a Nash equilibrium of the game?

These observations turn out to be no accident, for the following theorem can be proven:

Theorem 6 (Hofbauer and Sigmund, 2002) *Let G be a symmetric two-player game. Then*

1. *If σ is a Nash equilibrium of G, then σ, interpreted as distribution over a population of players following the pure strategies of the game G, is a fixed point of the replicator dynamics.*
2. *If $\vec{p}$ is a Lyapunov stable point of the game G under the replicator dynamics, then the corresponding mixed strategy is a Nash equilibrium of G.*
3. *If σ is an ESS of G, then σ, interpreted as a population distribution over pure strategies, is asymptotically stable under the replicator dynamics.*

None of the three conditionals of Theorem 6 hold in the converse. First, we have seen numerous examples of fixed points of the replicator dynamics which

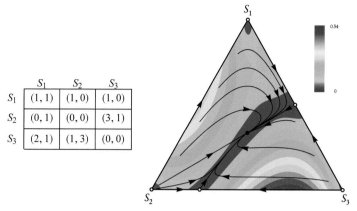

	S_1	S_2	S_3
S_1	$(1,1)$	$(1,0)$	$(1,0)$
S_2	$(0,1)$	$(0,0)$	$(3,1)$
S_3	$(2,1)$	$(1,3)$	$(0,0)$

Figure 16 An example of a case where, under the replicator dynamics, an asymptotically stable strategy does not correspond to an evolutionarily stable strategy of the underlying game.

don't correspond to Nash equilibrium (think All-Cooperate in the Prisoner's Dilemma). Second, there are Nash equilibrium strategies which, although fixed points, aren't Lyapunov stable (the 'bad' version of Rock-Paper-Scissors of Figure 15b). Third, and perhaps most surprising, though, is that not all asymptotically stable points of the replicator dynamics correspond to evolutionary stable strategies of the underlying game.

This last point is by no means immediately obvious. Zeeman (1979) was the first to provide an example of a case where an asymptotically stable strategy was not an ESS. Figure 16 shows another instance, using a game with a slightly simpler payoff structure. The asymptotically stable fixed point is $\vec{p} = \left\langle \frac{1}{3}, \frac{1}{3}, \frac{1}{3} \right\rangle$. Since all asymptotically stable points are also Lyapunov stable, it follows from Theorem 6 that the mixed strategy $\sigma = \frac{1}{3}S_1 + \frac{1}{3}S_2 + \frac{1}{3}S_3$ is also a Nash equilibrium of the underlying game, and so is a potential candidate for being an ESS. Now consider the mixed strategy $\mu = \frac{7}{16}S_1 + \frac{1}{4}S_2 + \frac{5}{16}S_3$. Because

$$\pi(\sigma \mid \sigma) = 1 = \pi(\mu \mid \sigma)$$

but

$$\pi(\mu \mid \mu) = \frac{131}{128} > \pi(\sigma \mid \mu) = \frac{49}{48}$$

it follows that σ is not an ESS, which shows the converse of the third statement of Theorem 6 is false.

Although it is true that an ESS of the underlying game will correspond to an asymptotically stable state of the replicator dynamics, that state may not be evolutionarily significant. Figure 17a shows one game which has an ESS: the strategy S_3. However, the size of the basin of attraction for S_3 is a function of

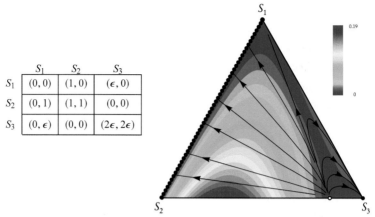

	S_1	S_2	S_3
S_1	$(0,0)$	$(1,0)$	$(\epsilon,0)$
S_2	$(0,1)$	$(1,1)$	$(0,0)$
S_3	$(0,\epsilon)$	$(0,0)$	$(2\epsilon,2\epsilon)$

(a) A game which illustrates both the preservation of weakly dominated strategies and that an asymptotically stable state which corresponds to an ESS of the underlying game may have an arbitrarily small basin of attraction. Every point on the S_1–S_2 edge is Lyapunov stable.

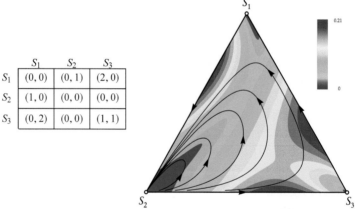

	S_1	S_2	S_3
S_1	$(0,0)$	$(0,1)$	$(2,0)$
S_2	$(1,0)$	$(0,0)$	$(0,0)$
S_3	$(0,2)$	$(0,0)$	$(1,1)$

(b) A game with three unstable fixed points, but where the basin of attraction for the All-S_2 state consists of the entire interior of the simplex.

Figure 17 Two games where the most likely outcome under the replicator dynamics will not be an ESS.

the parameter ϵ in the payoff matrix; as the value of ϵ approaches zero, the proportion of possible initial conditions which will converge to S_3 also approaches zero. This means that, even if we know that a particular strategy is an ESS (even *the* unique ESS), we cannot assume – in the absence of possessing any further knowledge about the population – that the ESS is likely to be present in the population to any significant extent. In the game of Figure 17a, if all we are told is that the initial state of the population was some initial condition selected at random, the safest bet, for small values of ϵ, is that *no one* would follow the ESS.

In Figure 17a, most of the population states converge to some fixed state containing a mix of S_1 and S_2. The precise mix of S_1 and S_2 depends upon the initial conditions. If the population starts off with not that many S_1-players, the final state will contain more S_2 than S_1; if the population starts off with a lot of S_1-players, the final state will contain more S_1 than S_2. But what is interesting is that, as a close look at the payoff matrix shows, S_2 is *weakly dominated* by S_1! In Section 2 we noted that a weakly dominated strategy could never be an ESS. Under the replicator dynamics, it is possible for weakly dominated strategies to survive and flourish, even driving an ESS to extinction.

While the replicator dynamics may not eliminate weakly dominated strategies, one can prove that strictly dominated strategies are always eliminated, in the limit:

Theorem 7 *Under the replicator dynamics, strictly dominated strategies are always driven to extinction as $t \rightarrow \infty$, provided that all pure strategies are initially present.*

The requirement that all pure strategies have to be present is needed because the replicator dynamics cannot add new strategies to the population. In the Prisoner's Dilemma of Figure 12a, the All-Cooperate state is a fixed point; however, it is an *unstable* fixed point because once any $\epsilon > 0$ of Defectors are included in the initial conditions of the population, Cooperate, as a strictly dominated strategy, is driven to extinction.

Finally, just to complicate things, the game of Figure 17b shows that sometimes a dynamically unstable fixed point might nevertheless be the evolutionarily expected outcome, in the long run. In that game, the all-S_2 state is not asymptotically stable (and, therefore, not an ESS) because any displacement which adds a small number of S_3-mutants will cause the population to move further away from the all-S_2 state. Over time, though, we see that the population will follow a circuitous route, with the proportion of S_3-players increasing, followed by the proportion of S_1 players increasing, and then eventually converging back to all-S_2 state in the limit. So, in this case we have a state which, in addition to not being an ESS, is *dynamically unstable*, but nevertheless is the long-term expected evolutionary outcome for almost every population state.

3.2.5 Chaos

At this point we have seen all of the possible types of evolutionary outcomes for games consisting of three strategies or fewer. A famous theorem from dynamical systems theory, the Poincaré–Benixson Theorem, states that there are only two possible outcomes for trajectories in continuous dynamical systems with a two-dimensional state space: either the trajectory converges to a fixed point, or the trajectory converges to a cycle. That does not mean that every trajectory must converge to the *same* fixed point or cycle, though. All of the games in Figure 12 have evolutionary trajectories which converge to a point, but the coordination game of Figure 12b has *two* different fixed points that trajectories can converge to, depending on where the population begins in state space. For the normal version of Rock-Paper-Scissors, almost every point in the state space lies on a cycle, and there are infinitely many cycles. For the 'bad' version of Rock-Paper-Scissors, almost every point in state space lies on a cycle which spirals outwards, converging towards the simplex boundary. Bomze (1983) provides a complete classification of all of the different possibilities, for three-strategy symmetric evolutionary games under the replicator dynamics, but the point is that, of the 48 different possibilities listed by Bomze, all consist of some combination of these two basic building blocks – trajectories which converge to a point, or to a cycle.

When we consider games with four or more strategies, more complex evolutionary outcomes become possible. But we first need to discuss how to visualise the evolutionary trajectories of games with four strategies. For such games, the state space is

$$ \left\{ \langle p_1, p_2, p_3, p_4 \rangle \in \mathbb{R}^4_+ \mid \text{where } p_1 + p_2 + p_3 + p_4 = 1 \right\}. $$

Figure 18a shows the set of all such points represented as a subset of Euclidean three-dimensional space. Although the games have four strategies, the state space is a three-dimensional object because once we have fixed the frequencies of three strategies, we know that the rest of the population must follow the remaining strategy. Although this is a perfectly accurate representation of the state space (no information is omitted), it does have the aesthetic peculiarity that not all faces of the polygon are the same. The face corresponding to the set of points where S_4 is missing from the population is different from the other three faces. For this reason, in the evolutionary game theory literature it more common to use the state space representation shown in Figure 18b. All we have done in moving from Figure 18a to Figure 18b is change the basis vectors used,[45] but we arrive at a more aethestically pleasing representation, where each face – representing a restricted version of the game where only

---

[45] In Figure 18a, the basis vectors were the standard basis of $\mathbb{R}^3$: $\vec{e}_1 = \langle 1, 0, 0 \rangle$, $\vec{e}_2 \langle 0, 1, 0 \rangle$, and $\vec{e}_3 = \langle 0, 0, 1 \rangle$. In Figure 18b, the basis used is $\vec{e}_1 = \langle 1, 0, 0 \rangle$, $\vec{e}_2 = \left\langle \frac{1}{2}, \frac{\sqrt{3}}{2} \arcsin\theta, \frac{\sqrt{3}}{2} \arccos\theta \right\rangle$, and $\vec{e}_3 = \left\langle \frac{1}{2}, \frac{\sqrt{3}}{2}, 0 \right\rangle$, where θ denotes the angle between faces of the regular tetrahedron (approximately 70.5°).

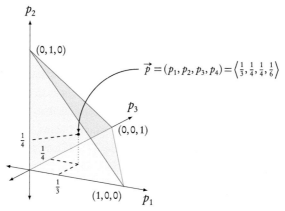

(a) The state space for a population playing a game with four strategies, represented as a subset of normal $\mathbb{R}^3$ Euclidean space. Since the population proportions must sum to one, the values of p_1, p_2, and p_3 uniquely determine the value of p_4. All possible population states occur within the shaded irregular polygonal region (including the faces). A point in the region represents a population state, as illustrated.

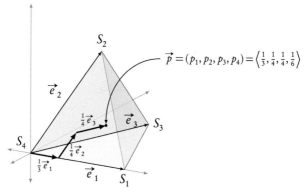

(b) A more common representation used for evolutionary games with four strategies. The region shown in Figure 18a is transformed into a regular tetrahedron. Each face, where only three strategies are present, is thus a normal simplex as previously used for three-strategy games. A point in the interior of the tetrahedron, representing a population state where all four strategies are present, is obtained by a linear combination of the three basis vectors $\vec{e}_1$, $\vec{e}_2$, and $\vec{e}_3$, weighted by the appropriate population proportion, as shown. Each vertex of the tetrahedron represents a population state where everyone follows the same strategy (as labelled).

Figure 18 Two equivalent representations of the state space for four-strategy evolutionary games.

three strategies are present – is the same type of triangle used before when representing the state space of three-strategy games.

In games with four or more strategies, *chaotic* behaviour becomes possible. First discovered by Edward Lorenz in 1963, chaos refers to behaviour in dynamical systems that appears 'random', 'chaotic', or 'unpredictable' *even though the system is deterministic*. A proper mathematical characterisation of chaos is well beyond the scope of this volume, but the intuitive idea begins with *sensitive dependence on initial conditions*.[46] In many systems we are familiar with, small differences in the starting state of the system either generally disappear over time,[47] or remain 'small' differences as the state of the system changes over time.[48] In chaotic systems, small differences in the initial conditions can yield such radically different future trajectories that, even if we know the complete future state $\vec{p}(t)$ of the system which begins at $\vec{p}(0)$, we will know very little about the complete future state $\vec{q}(t)$ of a system whose initial state $\vec{q}(0)$ is only ϵ away from $\vec{p}(0)$. As is often said: chaos is why long-term precise weather prediction is impossible.[49]

Figure 19 illustrates one chaotic evolutionary trajectory for a four-strategy game, originally identified by Arneodo et al. (1980). The particular version of the game shown here is due to Sandholm (2010).[50] Unlike all games seen previously, here the population converges neither to a point nor to a cycle. Instead, it follows a complicated intertwining path which never *exactly* returns to a previously visited state, but which at the same time is confined to a particular subregion of state space, known in chaos theory as a *strange attractor*. When an evolving population becomes trapped by a strange attractor, it will exhibit an fascinating kind of nonperiodic behavior. Let $\vec{p}_{t_0}$ denote the point in state space occupied by the population at time t_0, after it has become trapped by the strange attractor. Then there will be future times t_1 and t_2, with $t_0 < t_1 < t_2$, so that, at t_1, the population will be the point $\vec{p}_{t_1}$, arbitrarily far away from $\vec{p}_{t_0}$, and at t_2 the population will be at the point $\vec{p}_{t_2}$, arbitrarily close to $\vec{p}_{t_0}$. That is to say, once the population has been trapped by the strange attractor, although

---

[46] One common definition of chaos is due to Devaney (1989, p. 50), who requires, in addition to sensitive dependence on initial conditions, topological transitivity and dense periodic points. These three conditions are described by Devaney as covering 'unpredictability, indecomposability, and an element of regularity.' The last two conditions (topological transitivity and dense periodic points) are often omitted in popular discussions of chaos, but they are necessary because simple deterministic systems, like $f(x, t) = e^{xt}$, exhibit sensitive dependence on initial conditions without being chaotic, in the sense that $f(x + \epsilon, t) - f(x, t)$ will become arbitrarily large as t increases. In some dynamical systems, the latter two conditions actually imply sensitive dependence.

[47] Recall the two-strategy coordination problem from Figure 12b. Small differences in the initial conditions will eventually wash out over time, provided that the difference doesn't move the system across the unstable fixed point at $p_1 = \frac{1}{2}$ which divides the state space into two basins of attraction.

[48] In the normal version of Rock-Paper-Scissors, small differences will only shift the system to a slightly different cycle around the unstable fixed point in the centre of the simplex.

[49] Which is very different from why long-term *climate* prediction is possible. Even though we are unable to predict, at any point in time, what exact pattern of bubbles or Schlieren lines will be found in a pot of water on a hot stove (due to chaos), we know exactly what the end state will be (due to thermodynamics). It is going to boil, whether you watch it or not.

[50] Skyrms (1992, 1993) also has a discussion of chaos in the replicator dynamics.

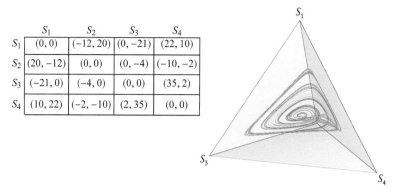

	S_1	S_2	S_3	S_4
S_1	$(0, 0)$	$(-12, 20)$	$(0, -21)$	$(22, 10)$
S_2	$(20, -12)$	$(0, 0)$	$(0, -4)$	$(-10, -2)$
S_3	$(-21, 0)$	$(-4, 0)$	$(0, 0)$	$(35, 2)$
S_4	$(10, 22)$	$(-2, -10)$	$(2, 35)$	$(0, 0)$

Figure 19 Chaos in the replicator dynamics. (Source: Reproduced with permission from Sandholm 2015, Copyright Elsevier.)

it will never leave that region of state space, its future behaviour will vary – nonperiodically – between moving as far away as you like from earlier states (subject to the constraint of remaining in the attractor) and moving as close as you like to earlier states (without ever *exactly* coinciding).

3.3 The Brown–Nash–von Neumann and Smith Dynamics

The replicator dynamic is the most commonly studied dynamic for evolutionary games, but other dynamics exist. In what follows, I shall briefly discuss two others: the Brown–Nash–von Neumann dynamic (Brown and von Neumann, 1950) and the Smith dynamic (Smith, 1984). Although less frequently studied, these two dynamics are interesting because the underlying belief revision processes which yield them are both plausible, in their own right, and arguably could be seen as rational improvements – from the individual's point of view – on the belief revision process which yields the replicator dynamic.

To begin, let us fix the game under consideration to be the 'bad' version of Rock-Paper-Scissors as in Figure 15b. But now suppose that the rate at which an individual decides to switch to Rock, Paper, or Scissors is independent of *both* their current strategy *and* their current expected payoff. Instead, suppose that the rate at which a player switches to a particular strategy S_j only depends on whether the expected fitness of S_j exceeds the average fitness of the population. For this belief revision process, we have

$$\rho_{ij} = \left[W_j - \overline{W} \right]_+ .$$

In a sense, this belief revision process is a rational improvement on the one which yields the replicator dynamics because it doesn't require that strategies need to be present in the population to be considered. Imitation, as a guide to behaviour, can mislead because the mere fact that a strategy does

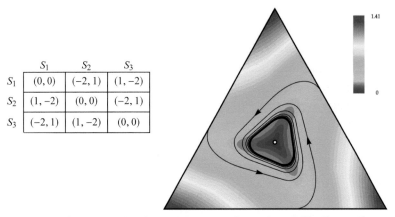

	S_1	S_2	S_3
S_1	$(0, 0)$	$(-2, 1)$	$(1, -2)$
S_2	$(1, -2)$	$(0, 0)$	$(-2, 1)$
S_3	$(-2, 1)$	$(1, -2)$	$(0, 0)$

Figure 20 The Brown–Nash–von Neumann dynamics, unlike the replicator dynamics, can introduce new strategies into the population.

well, given the current state of the population, doesn't mean that there isn't an alternative, unrepresented, strategy which would do better. Yet this belief revision process remains a boundedly rational one because individuals don't adopt a pure best response to the population state: any strategy whose expected fitness exceeds the average fitness of the population will have a positive adoption rate.[51] This belief revision process, at the population level, yields the *Brown–Nash–von Neumann* (BNN) dynamic.

As Figure 20 shows, the evolutionary behaviour for the 'bad' variant of Rock-Paper-Scissors is significantly different under the BNN dynamic than under the replicator dynamic. Each of the three trajectories shown begin in a state where everyone follows one of the three pure strategies. Unlike the replicator dynamics, the BNN dynamics can introduce new strategies into the population if they are not initially represented. When the population starts in the all-S_1 state, some S_2 followers are introduced. The number of S_2-players increases over time, resulting in states featuring more and more S_2 players, until a tipping point is crossed when there are so many S_2 players that

---

[51] This fact means the process being modelled is one of *biased* random mutation: any possible strategy whose expected fitness is better than the average fitness of the population has a nonzero adoption rate, but the adoption rate can vary based on the expected fitness of the strategy. There are models of *nonrandom* mutation in evolutionary games as well. Perhaps the most straightforward such model would involve a belief revision process based on best response. Although the idea behind such a dynamic is quite intuitive, the mathematical analysis proves to be considerably more complicated. Under a best-response dynamic, for example, solutions for the population trajectory from an initial starting point may not be unique. Those interested in the details about best-response dynamics for continuous dynamical models of evolutionary games should consult chapter 6 of Sandholm (2010).

following S_3 confers a greater expected fitness than the average of the population. At this point, the dynamics have moved the state of the population off the edge of the state space into the interior. From there, the dynamics will lead to a greater mix of all three strategies until it finally converges on a cycle.

One unusual aspect about the belief revision processes which yield the replicator dynamics or the BNN dynamics is that they compare how well a potential strategy does against the average fitness of the population. Most of the time, *there is no strategy* whose expected fitness equals the average fitness of the population, and so one might wonder why that is the natural benchmark for measuring a strategy's performance. For example, if one strategy has an absolutely *appalling* performance, it could pull down the average fitness so much that some strategies which would not normally be seen as desirable nevertheless are viewed as contenders. Given this, an alternative possibility was proposed by Smith (1984) who suggested that the rate of change between strategy S_i and S_j should only depend on whether the expected fitness of S_j was greater than that of S_i:

$$\rho_{ij} = \left[W_j - W_i \right]_+ .$$

As a boundedly rational belief revision process, Smith's proposal in some ways improves upon the one which yields the BNN dynamics. How so? By avoiding comparison with the average fitness of the population, Smith's proposal ensures that individual choices are based on fitness improvements attainable by actual strategies. In addition, just like with the BNN belief revision process, this process doesn't require a strategy be present in the population to be considered for adoption. When individual choices are aggregated, at the population level we obtain what is known as the *Smith dynamic*.

Like the BNN dynamic, the Smith dynamic allows for the introduction of new strategies not initially present in the population. Hofbauer and Sandholm (2011) also show that the Smith dynamic allows for a phenomenon which we know cannot happen under the replicator dynamics: the preservation of *strictly* dominated strategies. Figure 21 shows a four-strategy game known as 'Bad' Rock-Paper-Scissors with a feeble twin. The fourth strategy added to the typical 'Bad' Rock-Paper-Scissors game is similar to Scissors, except for the fact that its payoffs are always ϵ lower.

Under the Smith dynamic, populations beginning from pure states consisting of all one type introduce competing types where it is advantageous to do so. For example, from the pure Scissors state the population will add players following the strategy Rock and, once the proportion of Rock players becomes large enough, Paper-players will join the population as well. The surprising fact about the Smith dynamic is that because the adoption rate ρ_{PT} is positive

	R	P	S	T
R	$(0,0)$	$(-2,1)$	$(1,-2)$	$(1,-2-\epsilon)$
P	$(1,-2)$	$(0,0)$	$(-2,1)$	$(-2,1-\epsilon)$
S	$(-2,1)$	$(1,-2)$	$(0,0)$	$(0,-\epsilon)$
T	$(-2-\epsilon,1)$	$(1-\epsilon,-2)$	$(-\epsilon,0)$	$(-\epsilon,-\epsilon)$

(a) The game of 'Bad' Rock-Paper-Scissors with a feeble twin.

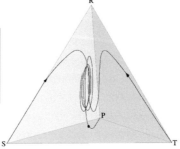

(b) The preservation of strictly dominated strategies under the Smith dynamic.

Figure 21 The Smith dynamic allows for the preservation of strictly dominated strategies.

(the Twin strategy does receive a positive payoff against Paper, even though the Twin's payoff is not as high as Scissors' payoff), for small values of ϵ the strictly dominated Twin strategy will persist indefinitely. The population will follow a cycle containing a mix of all four strategies.

Furthermore, the survival of strictly dominated strategies is not an isolated phenomenon peculiar to the Smith dynamics. One can prove a general result showing that *any* evolutionary dynamics which satisfy four general conditions will allow the preservation of strictly dominated strategies. These conditions are as follows:

Continuity: The function describing the growth rate of strategies in the population needs to satisfy a fairly standard continuity condition,[52] ensuring that small changes in population composition or payoffs do not result in huge changes in the dynamics.

Positive Correlation: If the composition of the population is changing, then the growth in the frequency of a strategy in the population and its payoff is positively correlated.

Nash Stationarity: The fixed points of the population dynamics only occur at Nash equilibria of the underlying game.

Innovation: When a population is not in a fixed state, and an unused optimal strategy exists, the growth rate of that strategy must be strictly positive.

Perhaps the most striking aspect of these four conditions is that they all seem fairly plausible ones to impose on an evolutionary dynamic – especially one which is supposed to represent cultural evolution. But, taken together, they imply the following:

---

[52] For those who have studied some mathematics: the function needs to be Lipschitz continuous.

Theorem 8 (Hofbauer and Sandholm, 2011) *Suppose the evolutionary dynamics satisfy Continuity, Positive Correlation, Nash Stationarity, and Innovation. Then there exists a game such that the evolutionary dynamics, beginning from most initial conditions, will have a strictly dominated strategy played by some strictly positive proportion of the population.*

It is important to note what Theorem 8 does *not* say. That theorem does not say that any game with a strictly dominated strategy will have that strategy preserved for most initial population states. Rather, the theorem shows (by construction) that there exists at least one game for which strictly dominated strategies are preserved from most initial population states.

Why does the replicator dynamics always eliminate strictly dominated strategies? Because the replicator dynamics is, in many respects, a peculiar dynamic from the point of view of rational actors, even boundedly rational ones. The replicator dynamics fails to satisfy Nash Stationarity and Innovation. It fails Nash Stationarity because not every fixed point of the replicator dynamics occurs at a Nash equilibrium of the underlying game. Recall the Prisoner's Dilemma: the All-Cooperate state is a fixed point – an unstable one, of course, but still a fixed point – even though (Cooperate, Cooperate) is not a Nash equilibrium of the game. It fails Innovation because of the fact that p_i occurs in the formula for the growth rate of S_i: if a strategy isn't present in the population, it cannot be introduced.

Of the dynamics considered in this section, both the BNN and the Smith dynamic satisfy the preceding four conditions, and thus both of them allow the preservation of strictly dominated strategies. From the point of view of modelling boundedly rational agents, it is somewhat surprising that the two belief revision processes which yield the BNN dynamic and the Smith dynamic both seem, at first blush, like more reasonable procedures to use than the one which yields the replicator dynamics. The replicator dynamics arises from imitative learning: individuals sample people from the population, compare them to the average, and update with probability proportional to success. Presumably this means that agents *don't know* the full payoff structure of the game, because it would be difficult to justify the assumption that players *did* know that structure while, at the same time, refusing to adopt a strictly dominant strategy unrepresented in the current population. Yet, although the processes which yield the BNN and Smith dynamic are improvements, from the point of view of the boundedly rational individual, they paradoxically lead to the preservation of strictly dominated strategies at the population level.

		Billy	
		Boxing	Ballet
Maggie	Boxing	(2, 1)	(0, 0)
	Ballet	(0, 0)	(1, 2)

Battle of the Sexes

Figure 22 The asymmetric game Battle-of-the-Sexes, first introduced by Luce and Raiffa (1957), where the payoff received at the two pure-strategy Nash equilibrium outcomes depends in part on the player's identity, rather than merely on the strategy adopted. (The payoff structure here assumes that the game is played between the respective leads of the films *Million Dollar Baby* and *Billy Elliot*.)

3.4 Multipopulation Games

All of the games discussed so far have been *symmetric* games: where the only thing that matters for the payoff received is the strategy used by a player. What the requirement of symmetry excludes are games like Battle of the Sexes, shown in Figure 22. In general, it's not possible to model asymmetric games using a *single* population without introducing additional assumptions that transform the strategic nature of the game. For example, if we assume that players are randomly assigned to be Row or Column (each role equally likely), then Battle of the Sexes becomes a simple pure coordination game.

Multipopulation models provide one way of analysing the evolutionary behaviour of asymmetric games. In a multipopulation model, we have separate populations whose individuals are always 'assigned', so to speak, the role of Row or Column player in a pairwise interaction.[53] We will refer to these as the Row (or Column) populations. These populations are assumed to evolve separately without there being any flow of individuals between them. What makes these evolutionary models interesting is that the *fitness* of individuals belonging to the Row population functionally depends on the composition of both the Row and Column populations. (The same is true for the fitness of individuals belonging to the Column population.) For example, consider a population of Row players for Battle of the Sexes, where $p_{Boxing} = \frac{1}{4}$ and $p_{Ballet} = \frac{3}{4}$. The fitness of the Boxing strategy is indeterminate until we specify the composition of the Column population. If a high enough proportion of Column players choose Boxing, then the frequency of Boxing will increase in the Row population and the system will converge on All-Choose-Boxing; however, if the proportion of Column players choosing Ballet is high enough, then Ballet will increase in the Row population and the system will converge on All-Choose-Ballet.

---

[53] In addition, there is no reason to restrict ourselves to just two player games. The current framework could easily be extended to N-player games, where we have N different subpopulations, with the ith player drawn at random from the ith population.

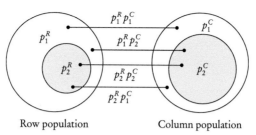

Row population   Column population

Figure 23 A two-population model for a two-strategy game. The respective probabilities of each type of interaction are indicated.

For a two-population model, the replicator dynamics takes the form of a coupled system of differential equations, as follows:

$$\frac{dp_i^R}{dt} = p_i^R \cdot \left[W_i^R \left(\vec{p}^C \right) - \overline{W^R} \right] \tag{1}$$

and

$$\frac{dp_i^C}{dt} = p_i^C \cdot \left[W_i^C \left(\vec{p}^R \right) - \overline{W^C} \right]. \tag{2}$$

Note that the functional dependence of W_i^R and W_i^C on the state of the *other* population has been explicitly indicated: the argument to W_i^C is $\vec{p}^R$ and the argument to W_i^R is $\vec{p}^C$.

Figure 24 illustrates some evolutionary trajectories for the two-population model of Battle-of-the-Sexes under the replicator dynamics. The state space appears different from previous diagrams because, although we have two independent populations, there are only two degrees of freedom: once we fix the proportion of the Row population which play Boxing (or Ballet), we know the remainder must play Ballet (or Boxing), and the same is true for the Column population. The diagram has been drawn so as to spatially align with the payoff matrix from Figure 22: the upper-left corner of the diagram represents the pure population states where both Row and Column consist of 100 per cent Boxing-strategists, and the lower-right corner represents the pure population states where both Row and Column consist of 100 per cent Ballet-players.[54] The point corresponding to the Nash equilibrium in mixed strategies, $\sigma_R = \frac{2}{3}$Boxing $+ \frac{1}{3}$Ballet and $\sigma_C = \frac{1}{3}$Boxing $+ \frac{2}{3}$Ballet is an unstable fixed point of a type known as a *saddle point*. If the Row and Column population

---

[54] There is one slight peculiarity which results from this representation: if you think of the diagram as part of normal Cartesian space, with the origin at the lower-left and the length of each side set to 1, the *x*-axis represents the proportion of *Ballet-strategists* in the *Column* population, increasing from 0 to 1. However, the *y*-axis represents the proportion of *Boxing-strategists* in the *Row* population, increasing from 0 to 1.

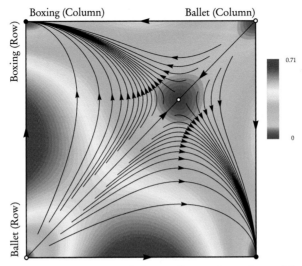

Figure 24 The two-population replicator dynamic model of
Battle-of-the-Sexes.

states happen to be *exactly* of the right relative proportions, then both populations will converge to the state corresponding to their Nash equilibrium mixed
strategy. However, if there is any misalignment whatsoever, the two populations will converge to states corresponding to one of the two pure strategy Nash
equilibria.

Much more could be said about multipopulation models. One fascinating
illustration of their use in philosophical decision theory can be found in *The
Dynamics of Rational Deliberation* (Skyrms, 1990). In that book, Skyrms
shows how the mathematical formalism of multipopulation models can be
interpreted as representing changes in the *degrees of belief* for two rational
agents facing an interdependent decision problem. According to that interpretation, what Figure 24 illustrates is how rational deliberation will lead to changes
in the degree of belief for two *individuals* facing a one-shot Battle-of-the-Sexes
game. What we have previously been considering as the 'initial conditions
of the population' are, instead, viewed as the initial *credences* of each player
regarding what to do.

Perhaps this is a fitting place to bring the discussion of this section to a close.
We began by showing how interactions between large numbers of individuals
gave rise to deterministic dynamical laws, at the level of the population, in
the form of the replicator dynamics. We then saw how attending to the evolutionary dynamics of populations led to a decoupling of notions of evolutionary

stability between the static analysis given in Section 2 and the dynamic analysis given here. We then saw how different belief revision procedures used by individuals gave rise to different dynamical laws in the BNN and Smith dynamics, and how these dynamics led to 'irrational' outcomes by allowing strictly dominated strategies to be preserved, even though the individual belief revision processes were ostensibly more rational than those which gave rise to the replicator dynamics. And, finally, we have come full circle in seeing how multipopulation evolutionary models can be reinterpreted as describing the rational deliberation of persons.

One conceptual point which we have not properly unaddressed is the fact that all of these dynamical models essentially assume infinite populations, where all pairwise interactions are equally likely. The real world is finite and small. Most people only interact with a few individuals they know. Does that make a difference? That is the topic of the next section.

4 Finite Population Models

In the previous section, we looked at a number of evolutionary models where the population state was represented by a vector $\vec{p}$ of positive real numbers which sum to 1. All of the models considered involved processes of *continuous* change, where dynamical laws operating at the level of the population specified the instantaneous rate of change given the composition of the population at a moment in time. In these continuous models, arbitrarily small changes could happen, and quite frequently made a material difference to the outcome, no matter how small. For example, the All-Cooperate state of the Prisoner's Dilemma is a fixed point of the replicator dynamics, but if *some* $\epsilon > 0$ of the population mutate to Defectors, no matter how small, the entire population will converge upon the All-Defect state. And this happens even if $\epsilon < 10^{-82}$, which is less than one divided by the total number of all the atoms in the universe.

In real life, populations are *discrete* rather than continuous. They consist of a finite number of individuals, with a varying population size, and an upper limit on how large the population can get. (The upper limit might not be known, and the limit might itself vary based on environmental factors; nevertheless, an upper limit certainly exists at all times.) And some populations – although not all – are *structured*, so that the possible interactions between individuals are restricted in some fashion. In this section we will look at how taking these factors into consideration can make an enormous difference in the evolutionary outcomes.

	A	B
A	(a,a)	(b,c)
B	(c,b)	(d,d)

Figure 25 The generic payoff matrix used for the birth-death process. It is assumed that $a, b, c, d > 0$.

4.1 The Birth-Death Process

The simplest evolutionary game theoretic model for a finite population is the *birth-death process*.[55] In this model, we assume that the population has a fixed size N with two possible strategies, which we will denote by A and B, rather than S_1 and S_2 as has been the convention up to now. (The change allows for a useful notational mnemonic, which we will see in a moment.) There is nothing special about the fact that the model only includes two types of individual: it could readily be generalised to allow for more types, but the mathematical analysis would then become more complicated. It is also assumed that there is a two-player, two-strategy symmetric game G, whose payoffs generate the expected fitnesses of each player. In order to avoid certain mathematical complications (like dividing by zero), we assume that the payoffs in G are strictly positive. This can always be achieved by transforming the payoffs by adding a common positive constant to every entry in the payoff matrix, as that does not alter the strategic form of the game. The payoff matrix then takes the form shown in Figure 25.

If we suppose that i players in the population follow strategy A, and $N{-}i$ players follow strategy B, and that players interact with someone selected at random, the expected *payoffs* are as follows:

$$W_i^A = a \cdot \frac{i-1}{N-1} + b \cdot \frac{N-i}{N-1}$$
$$W_i^B = c \cdot \frac{i}{N-1} + d \cdot \frac{N-i-1}{N-1}.$$

The above fitness calculations assume that all individuals are equally likely to be selected for interaction, so the chance of selecting an A-strategist is $\frac{i}{N}$ and the change of selecting a B-strategist is $\frac{N-i}{N}$. However, individuals cannot interact with themselves, which is why the denominators of both W_A and W_B are $N{-}1$,

---

[55] Some books on evolutionary dynamics begin with an even simpler model known as the *Moran process* (so called because it was first introduced by Moran, 1958). The Moran process is very similar to the birth-death process except for the fact that it assumes that there are no differences in the fitnesses between types. As such, the Moran process models *neutral* evolution and is of little interest for evolutionary game theory. Except, of course, for limit cases such as the game of life on page 38.

and the numerators which correspond to the payoffs generated by an $A \bullet\!\!-\!\!\bullet A$ or a $B \bullet\!\!-\!\!\bullet B$ interaction include factors of $i-1$ or $N-i-1$, respectively.

In the birth-death process, reproduction takes place in two stages: first, an individual is selected to *reproduce*, generating a clone; second, an individual from the 'old' population (not including the newly added clone) is selected to be *replaced*. This keeps the population at a constant size N. If a type is selected with a probability proportional to its fitness, then the probability of A or B being selected to reproduce in a population with i individuals following A is

$$p_i^A = \frac{iW_i^A}{iW_i^A + (N-i)W_i^B}$$

$$p_i^B = 1-p_i^A = \frac{(N-i)W_i^B}{iW_i^A + (N-i)W_i^B}.$$

If we assume that the chance of an individual being replaced is independent of their fitness,[56] we can then write down the *transition probabilities* for the two cases of interest: the population *increasing* in the number of A-types from i to $i+1$, which we will denote by α_i, or the population *decreasing* in the number of A types from i to $i-1$, which we will denote by β_i. These probabilities are

$$\alpha_i = p_i^A \cdot \frac{N-i}{N} = \frac{iW_i^A}{iW_i^A + (N-i)W_i^B} \cdot \frac{N-i}{N}$$

and

$$\beta_i = p_i^B \cdot \frac{i}{N} = \frac{(N-i)W_i^B}{iW_i^A + (N-i)W_i^B} \cdot \frac{i}{N}$$

In both cases, the transition probability consists of two components: that the new offspring will be of the relevant type (A or B), and that the individual replaced is of the *opposite* type. Of course, there is always the chance that the new offspring will be of the same type as the individual randomly selected for replacement. The probability of this happening is simply $1-\alpha_i-\beta_i$.

The birth-death process is a special type of random process known as a *Markov chain* (or a Markov process). In a Markov chain, the probability of the system being in a future state depends only on the current state of the system: there is no possibility of path-dependency. Figure 26 shows the state diagram for the birth-death process along with the transition probabilities. Each shaded

---

[56] Which is really not as implausible as it might sound. Think of it as the assumption that the chance of dying, from whatever cause, is constant across the population. Differences in fitness then correspond to the differential chance of certain types being included amongst the replacement progeny.

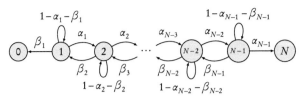

Figure 26 The state diagram for the Markov chain underlying a general birth-death process. Each shaded circle represents the population state containing i individuals following strategy A, for $i = 0, 1, \ldots, N$.

circle represents the state of the population with the number of A-type individuals displayed within. Given the reproductive dynamics of the model, there are two *absorbing states* which, once the system enters, it will never leave. These are the two end-states: one with no individual following strategy A and one with every individual following strategy A.

Given the nature of the birth-death process, any population consisting of some mix of As and Bs has a chance of ending up at either absorbing state. One interesting question is how likely it is that a population consisting *mostly* of the A-type (or B-type) can be displaced by a single mutant. If we consider the situation where we have a population of $N{-}1$ individuals following B and one individual following A, let ρ_A denote the *fixation probability* that A succeeds in taking over the entire population. Similarly, let ρ_B denote the fixation probability for B.

Nowak (2006) shows, through some clever algebraic manipulations, that an exact solution to the fixation probabilities is

$$\rho_A = \frac{1}{1 + \sum_{j=1}^{N-1} \prod_{k=1}^{j} \frac{W_k^B}{W_k^A}}$$

$$\rho_B = \frac{\prod_{k=1}^{N-1} \frac{W_k^B}{W_k^A}}{1 + \sum_{j=1}^{N-1} \prod_{k=1}^{j} \frac{W_k^B}{W_k^A}}.$$

To give a concrete example, consider the Prisoner's Dilemma defined by setting the values of the payoff matrix in Figure 25 to $a = 3$, $b = 1$, $c = 4$, and $d = 2$. As the value of N increases, numerical calculations of the fixation probability ρ_B (the chance of Defect taking over from a single mutant) show a clearly evident pattern:

	Population size, $N =$				
	10	100	1,000	10,000	100,000
ρ_B	0.34263	0.25879	0.25088	0.25009	0.25001

In fact, it is possible to show that $\rho_B \rightarrow \frac{1}{4}$ as $N \rightarrow \infty$.[57] This means that the replicator dynamics cannot be the limit of the birth-death process as the population size increases without bound. Why? In the birth-death process, the probability of Defect taking over converges to $\frac{1}{4}$ as the proportion of Defectors in the population converges to zero. But, under the replicator dynamics, the probability of Defect taking over the population is *always 1* for any $\epsilon > 0$ of Defectors in the population.

4.2 Local Interaction Models

One assumption which has persisted throughout all of the evolutionary models so far is that pairwise interactions have occurred *at random*, where the chance of a player interacting with a particular strategy is simply the proportion of that strategy in the population. While this is a reasonable assumption to make in certain contexts, it isn't true in general. When we consider human beings, people's interactions are heavily constrained by their respective social networks: you are much more likely to interact with your friends, family, and colleagues than a randomly selected person on the street. And even though technology has effectively decoupled our social networks from our physical location, those networks – a virtual map of who matters and in what way – still strongly influence our interactions and behaviours.

Such considerations are also found to matter amongst the simplest of biological organisms. In one experiment, Chao and Levin (1981) examine the outcome between two competing strains of the single-cell bacterium *E. coli*. One strain – call it the *poisoner* – produces colicin, a type of protein toxic to some forms of *E. coli*, but not toxic to the poisoner. You might expect the poisoner strain to drive out the strain susceptible to the poison, but there's a twist to the story. Manufacturing colicin causes the poisoners to reproduce more slowly, since it requires energy and resources. When poisoners and non-poisoners are mixed in a liquid culture, although the poisoner does free up some

---

[57] This is most easily done with a symbolic manipulation program like *Mathematica*. For the specified values of the Prisoner's Dilemma, the ratio $\frac{W_k^B}{W_k^A}$ simplifies to $\frac{2(N+k-1)}{N+2k-3}$. Substituting that into the formula for ρ_B and calling `FullSimplify` results in

$$\rho_B = \frac{3(N+3)\Gamma\left(\frac{N+1}{2}\right)\Gamma(2N-1)}{6\Gamma(2N)\Gamma\left(\frac{N+1}{2}\right) + (3-N)\Gamma(N)\Gamma\left(\frac{3N}{2}-\frac{1}{2}\right)}$$

where $\Gamma(x)$ denotes the standard gamma function from complex analysis. (The gamma function is an extention of the factorial $n! = 1 \times 2 \times \cdots \times (n-1) \times n$ to most of the complex plane.) The value of ρ_B is the closed-form solution giving the fixation probability for any particular population size N. Taking the `DiscreteLimit` as $N \rightarrow \infty$ yields $\frac{1}{4}$.

extra resources by killing the susceptible strain, in a mixed environment those extra resources are shared equally by both strains. Thus the lower growth rate of the poisoners actually places them at a selective disadvantage against the susceptible strain, and the poisoners are driven out. However, if the two competing strains of bacteria are cultivated on a flat plate of agar (a thick, jelly-like substance), Chao and Levin found that the poisoners were able to thrive. Why? Because the poison killed susceptible bacteria in the immediate vicinity of the poisoners, who were then able to take full advantage of the freed-up resources. In this case, *local interaction* effects, deriving from the physical proximity of bacteria on the plate of agar, made a real difference in the evolutionary outcome.

A local interaction model of evolutionary games examines what happens when individual interactions are constrained according to some kind of underlying structure. The structure may derive from spatial proximity, as with our poisoner bacteria, or it may derive from other kinds of structures, like friendship networks. To begin, assume we have a two-player symmetric game Γ with $S = \{S_1, \ldots, S_k\}$ denoting the set of strategies and π the common payoff function for all players. Let $V = \{i_1, \ldots, i_n\}$ be the set of players. Why 'V'? Because we will represent each player as a *vertex* in a *graph*.[58] Let E be the set of *edges* connecting the players, where each edge $e \in E$ is a set $\{i_j, i_m\}$, where $i_j, i_m \in V$. The set of edges E represents the possible pairwise interactions between players: if an edge connects two players, they can interact, and if no edge connects the players, they cannot. Two players connected by an edge are said to be *neighbours*, and the set of all individuals connected to a player i is the *neighbourhood* of i. The set of vertices and edges taken together defines a graph $G = (V, E)$. Since each edge is a set of players, with no natural ordering, we are using what is known as an *undirected* graph to represent the structure of this evolutionary model.

Given how complex the structure of graphs can be, it may not always be obvious how best to represent a graph's structure on a two-dimensional page. Figures 27a and 27b show two different representations of the same graph – a "scale-free" network generated using the Barabási and Albert (1999) preferential attachment model.[59] The circular embedding on the left does show that one vertex lies on a large number of edges, but the overall structure (indeed, even

---

[58] Here we are using the term 'graph' as in the branch of mathematics known as *graph theory*, rather than in the sense of 'the graph of the function $f(x) = 2x^2 + 1$'.

[59] This is a model of graph formation where new vertices, each having a fixed number of edges k, are added to an existing graph. The new vertex is connected to k other vertices at random, where the probability of connecting to a vertex is proportional to how many *other* vertices are already connected to it. In short, more 'popular' vertices are more likely to be chosen.

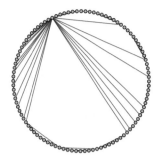

(a) A graph with 100 nodes, generated according to the Barabási and Albert (1999) model of preferential attachment with $k = 1$, shown with a circular embedding.

(b) The same graph as on the left, but drawn using a spring-electrical embedding. Nodes are treated as electrical charges (which mutually repel) and edges are treated as springs (which attract the two nodes they connect). The final layout is a minimal-energy configuration.

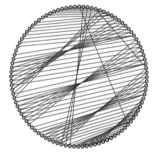

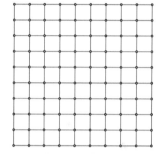

(c) A 10×10 grid graph, shown with a circular embedding.

(d) The same graph as on the left, but drawn using the standard spatial embedding for a grid graph.

Figure 27 Two different graphs, each represented in two different ways. The standard circular embedding, shown in (a) and (c), is the most simple visualisation but it often fails to reveal important aspects of the graph's structure. The two embeddings shown in (b) and (d) provide a more useful visualisation of each graph's structure.

the fact that the graph is connected) is otherwise obscured. Figures 27c and 27d illustrate the same point: the circular embedding on the left suggests an underlying *regularity* of the graph's structure,[60] but the fact that it is of a basic grid structure is also obscured. The point is this: since all that matters is whether two vertices are connected by an edge, the actual position of the vertices on the page doesn't matter, and so we can position them as we wish in order to make the structure more clear.[61]

---

[60] Note, though, that even this perceived regularity could be hidden if one used a different circular embedding – say, one that used a randomly generated reordering of the vertices around the circle.

[61] In this sense, graph representations are unlike geological *maps*.

One of the advantages of local interaction models of evolutionary games is that their discrete nature allows a very fine-grain level of control over modelling the evolutionary process. Unfortunately, this also means that almost every modelling assumption has plausible alternatives that could equally well be explored. As such, this means that one must exercise caution when interpreting the outcome of the model, because it may not always be clear which assumptions are crucial for driving the results. Occasionally – as we will see – apparently innocent modelling assumptions can turn out to make a big difference.

To begin, let us assume that each player is assigned a strategy and that each player uses the *same* strategy for all of their interactions – at least, that is, until they change it. But if a player changes their strategy, they continue to use the new strategy for *all* of their interactions until they change their strategy again. What this assumption excludes is the possibility that a player may knowingly *condition* their play based on the *identity* of the person they are interacting with.

Is that a plausible assumption? In some cases, it is. Poisoner *E. coli* bacteria, living on an agar plate, target all bacteria around them rather than specific neighbours. In the case of human players, whether the assumption is plausible depends on the situation. For the perfectly rational agents frequently studied by traditional game theorists and economists, such an assumption would seem implausible: if you have information about whom you are interacting with, why would you fail to include that in your deliberation? But for imperfectly rational agents – agents who "satisfice"[62] rather than optimise, or agents who make decisions using fast-and-frugal heuristics[63] – the assumption may not be unreasonable. If calculating the best response for each individual interaction

---

[62] This terminology derives from the work of Simon (1955, 1956) who suggested that, in many areas of life, people try to take decisions which are *satisfactory* rather than *optimal*. Such agents are said to engage in *satisficing* behaviour rather than *optimising* behaviour. For example, if you are trying to sell something – like a house – instead of trying to extract the highest possible price from the market, you might have a private aspiration level and you will accept any offer which exceeds that. (Simon develops this example in an appendix to his 1955 paper.) Since Simon first introduced the term 'satisficing' in his 1956 paper, there has been some conceptual drift in the understanding of the term. Artinger et al. (2022) identify two different senses of "satisficing" which have been used in the economics literature over the years, leading to two divergent research programmes.

[63] An alternative approach to less-than-perfectly-rational agents is due to Gigerenzer et al. (1999), who argues that people often make decisions using quick decision rules which are not guaranteed to deliver the best outcome, but which take advantage of features of the choice environment to deliver pretty good outcomes. For example, if you are on a quiz show and are asked which of two cities is the largest, and you have only heard of one, guessing that the city you've heard of is the answer is not a bad strategy.

is expensive or informationally impossible, using the same strategy against all opponents may make sense.[64]

In addition, let us assume that the evolutionary process of the model proceeds in two stages. In the first stage, every individual plays the underlying base game Γ with everyone they are connected to via an edge. People who are connected to many individuals therefore play the same game multiple times, receiving an aggregate payoff equalling the sum of the outcomes from each individual game. Depending on the structure of the underlying graph, this may mean that not all players have the same number of interactions in a given round of play. If we think of the graph as a social network, this makes sense: some people have a lot of friends and as a result have a number of social interactions every single day, whereas people with fewer friends have fewer interactions. In the second stage, after everyone has played the game and received an aggregate payoff, people potentially change their strategy using a *learning rule*. Learning rules are similar to the revision protocols of Sandholm (2010) in that they use information from the environment to determine whether a player will change their strategy. One reason for not adopting the term 'revision protocol,' here is to avoid confusion because the information available to a learning rule differs significantly from the information used by Sandholm's revision protocols. In the discussion of revision protocols on page 37, we saw that they use $\vec{W}$ and $\vec{p}$ to determine the player's next choice of strategy. In general, this *global* information about the population is not available to players in *local* interaction models. Instead, each player uses the information about the strategies and payoffs received from everyone in their neighbourhood to figure out what strategy to use in the next round of play.[65] Once everyone has reflected on whether to change their strategy – which doesn't necessarily mean that they will change their strategy – the process starts again at the first stage.

At this point we can already start to appreciate the earlier remark regarding alternative modelling assumptions. Even staying within the simple framework described earlier, a number of questions naturally arise: What kinds of graphs will determine the local interactions? What learning rules will be considered?

---

[64] This is one example of the fine-grain level of control over the details of modelling the evolutionary process. One can easily imagine alternative models where this assumption is relaxed. For example, Vanderschraaf (2007) develops a model of the Hobbesian state of nature where reputational information about the players can spread, and people can condition their choice of strategy on a person's reputation.

[65] A learning rule doesn't need to use *all* of the information it is given. For example, the *naïve best-response* learning rule assumes that everyone in the neighbourhood of a player will continue to use the same strategy in the next round of play, and then determines the best response using the payoff function for the underlying game, ignoring the information about the actual payoffs received by people in the current round of play.

Why is the neighbourhood people learn from necessarily the same as the neighbourhood people interact with? (The preceding definition indicates that it is the same; however, it is also obvious that it need not be.) Will all individuals use the same learning rule, or will the population consist of multiple types? Why is it assumed that the evolutionary process can be neatly split into an interaction stage followed by a learning stage? Can people make mistakes (or, can mutations insert new strategies into the population)? And so on.

Those questions only pertain to the formal structure as defined earlier – questions also arise which push us to go beyond the conception of local interaction models as defined previously. For example, why is the underlying graph structure treated as fixed rather than dynamic? Why is the underlying game assumed to be symmetric? Why aren't individuals allowed to condition their play on the identity of individuals? Why aren't players' strategies more sophisticated, e.g., having memory of the past plays of people within one's interaction neighbourhood so as to take that into account? All of these questions are perfectly reasonable to ask and have been studied by a number of people.[66] However, for the purpose of starting with simple, comprehensible models, we begin with the assumptions outlined earlier and then consider more complex models later. In what follows, we will consider several games which have been studied in the literature to varying degrees, showing how population structure can affect the evolutionary outcome.

4.2.1 Rock-Paper-Scissors

Figure 28 illustrates the outcome for a simple local interaction model of the standard version of Rock-Paper-Scissors played on a ring of nine people.[67] Each player has two neighbours, and the population consists of three clusters of people following the strategies Rock, Paper, or Scissors. The strategies followed by each player are shown inside the circle representing the player, and the payoffs received by each player are shown in the shaded gray square adjacent to the player. If players revise strategies using *imitiate-the-best*, we see that the pattern of strategies throughout the population has rotated clockwise by one person. Given that a clockwise rotation changes nothing about the strategic situation faced by (a) the person in the interior of a strategy-cluster (i.e., the Rock-player with two other Rock-players as neighbours), or (b) the persons on the periphery of a strategy-cluster (i.e., the Rock-player whose neighbours

---

[66] See, for example, Vanderschraaf (2006), Smead (2008), and Spiekermann (2009).

[67] This local interaction model has a very simple network structure, but it is more general than might initially appear. Suppose we have a room containing N people where each person has *exactly* two friends in the room. Then the friendship network of the room will either consist of a single ring of size N or else it will consist of k rings of size $N_1, \ldots, N_k$ where $N_1 + \cdots + N_k = N$.

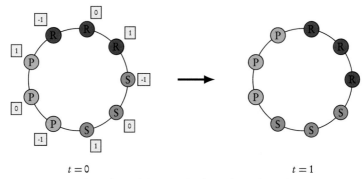

$t = 0$ $t = 1$

Figure 28 One iteration of the standard version of Rock-Paper-Scissors
played on a ring.

are one Paper-player and one Rock-player), from this we know that the population will cycle forever, each strategy-cluster rotating one person clockwise
after each generation.

On a one-dimensional ring in the absence of mutation, there are only two
possible evolutionary outcomes: cycling behaviour, or the entire population
eventually adopting the same strategy. (The latter outcome is most likely
to happen in small populations, where the initial random allocation of strategies doesn't result in the configuration required for cycles.) What may vary
are (a) the size of the strategy clusters, (b) whether the cycles go clockwise
or counter-clockwise, and (c) whether there are several Rock-Paper-Scissors
regions present on the ring.

When we consider the game of Rock-Paper-Scissors played in two-dimensional space, more complex behaviour becomes possible. Figure 29 shows
one outcome for a 30 × 30 lattice, where individuals interact with their four
neighbours at the cardinal compass points: North, South, West, and East. The
underlying graph structure is that of Figure 27d, with the exception that it wraps
at the edges: players on the left are connected to the corresponding player on
the right, and players on the top are connected to the corresponding player on
the bottom. Figure 29a shows the initial condition of the population with the
edges connecting players suppressed for clarity. After one hundred rounds of
play, we see that the population has settled into a cycle of period 12: regions
of players following Rock are eventually replaced by players following Paper,
which are then eventually replaced by players following Scissors, which then
return to players following Rock. The two-dimensional lattice model shows
evolutionary behaviour reminiscent of that of *Uta stansburiana*, the common side-blotched lizard whose behaviour we discussed at the start of this
Element.

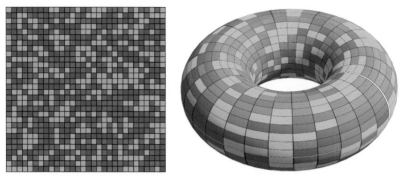

(a) The initial state of the population at $t = 0$. The image on the left doesn't show that the graph structure wraps at the edges. The actual graph structure is that of a torus, as shown on the right. However, because the three-dimensional toroidal representation hides some of the strategies, the convention is to show the population state as a flat image with the understanding that the edge behaviour is understood.

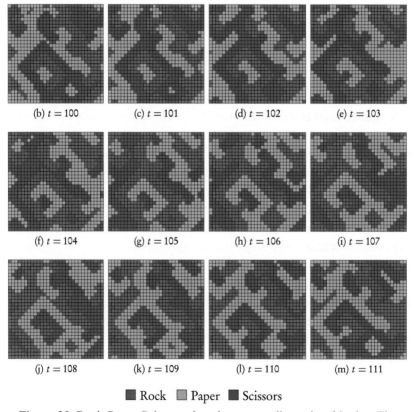

(b) $t = 100$ (c) $t = 101$ (d) $t = 102$ (e) $t = 103$

(f) $t = 104$ (g) $t = 105$ (h) $t = 106$ (i) $t = 107$

(j) $t = 108$ (k) $t = 109$ (l) $t = 110$ (m) $t = 111$

■ Rock ▢ Paper ■ Scissors

Figure 29 Rock-Paper-Scissors played on a two-dimensional lattice. The lattice wraps at the edges, so that the overall topology is that of a torus, as shown.

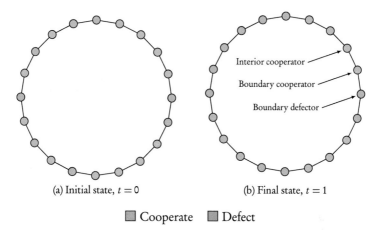

(a) Initial state, $t = 0$ (b) Final state, $t = 1$

☐ Cooperate   ☐ Defect

Figure 30 Groups of cooperators, if they are large enough, can survive in local interaction models. This figure shows two stable groups of cooperators for the Prisoner's Dilemma played on a ring. (Payoff matrix: $T = 1.1, R = 1.0$, $P = 0, S = -0.1$.)

4.2.2 The Prisoner's Dilemma

Turning attention to the Prisoner's Dilemma, we find that local interaction models allow for completely different evolutionary behaviour than what we saw under the replicator dynamics. Figure 30 shows that, on a ring where each player interacts with their immediate left and right neighbours, the strategy of Cooperate can persist indefinitely if the local cluster of cooperators is large enough and the payoff structure of the game has a certain form. In this case, the reason why Defect is not able to completely eliminate Cooperate is because players in the interior of the Cooperate region only interact with other cooperators, receiving, in this case, a payoff of 2. A Defector on the boundary of the cooperative region receives a payoff of 1.1 and, although this is greater than the payoff of 0.9 received by the boundary cooperator, the interior cooperator has the highest score of all the neighbours of the boundary cooperator. Under *imitate-the-best*, the boundary cooperator will "adopt" the strategy of Cooperate. And so we see how a region of cooperators can be stable, if it is large enough. But it is important to note that whether or not Cooperate can form stable regions depends on the payoff matrix. If the payoffs were $T = 1.5, R = 1.0$, $P = 0.6$, and $S = 0$, then the boundary defector would have the highest payoff of all the neighbours of the boundary cooperator, and no Cooperate region, no matter how large, would be stable for the simple ring.

Figure 31 illustrates how, when the interaction neighbourhood consists of the *two* closest individuals on both the left and right of a ring (a graph structure

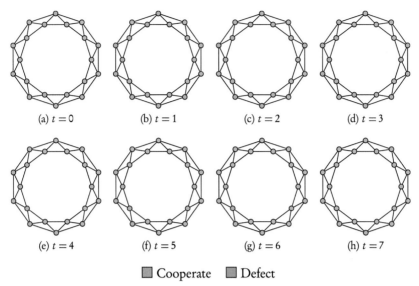

(a) $t = 0$ (b) $t = 1$ (c) $t = 2$ (d) $t = 3$

(e) $t = 4$ (f) $t = 5$ (g) $t = 6$ (h) $t = 7$

■ Cooperate   ■ Defect

Figure 31 When the interaction neighbourhood consists of the two closest players on the left and right, the strategy of Cooperate can drive out Defect. After seven iterations, the population enters into a cycle with period 3. (Payoff matrix: $T = 1.1$, $R = 1.0$, $P = 0$, $S = -0.1$.)

also known as a $1, 2$-circulant graph),[68] the strategy of Cooperate can not only *persist* but can *drive out* Defectors. This happens for two reasons. First, the random initial conditions happen to create a cluster of five adjacent Cooperators. Second, one of the Cooperators who has the boundary Defector as a neighbour interacts with three other Cooperators, receiving a payoff of 3. The boundary Defector, who has two Cooperators as neighbours, only receives a payoff of 2.2,

---

[68] An $n_1, n_2, \ldots, n_k$-*circulant graph* with N nodes is defined as follows: first, arrange the N nodes in a circle. Then connect each node to the two nodes located n_1 steps away counting to the left and right of the circle. Then, connect each node to the two nodes located n_2 steps away, again counting to the left and right of the circle. Continue in this fashion until you connect each node to the two nodes located n_k steps away on the left and right. For example, a $2, 4$-circulant graph of size 10 looks like this:

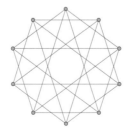

and so switches to Cooperate under *imitate-the-best*. Over time, the region of Cooperate will expand until it manages to reduce the number of Defectors down to a single player. At that point, the lone Defector will receive a payoff of 4.4, the highest possible payoff, and all of her adjacent Cooperators will switch to Defect, creating a region of five adjacent Defectors. However, this region of Defectors is not stable and will shrink over the next two iterations until, once again, we end up with a single Defector surrounded by Cooperators. This periodic behaviour will repeat indefinitely.

Understanding the evolutionary behaviour of models on simple structures such as cycles or circulant graphs is useful because, in some cases, it can help us figure out the possible evolutionary outcomes for other structures. One example where this can be done concerns a type of graph known as a 'small-world network' (see Figure 32a for an illustration).

Small-world networks derive their name from an early experiment conducted by the psychologist Stanley Milgram,[69] which explored the structure of the social network which enable two randomly chosen people to be connected through a chain of acquaintances. Milgram begins his article as follows:

> Almost all of us have had the experience of encountering someone far from home who, to our surprise, turns out to share a mutual acquaintance with us. This kind of experience occurs with sufficient frequency so that our language even provides a cliché to be uttered at the appropriate moment of recognizing mutual acquaintances. We say, 'My, it's a small world.' (Milgram, 1967, p. 61)

Milgram asked randomly selected people from two Midwestern towns (Wichita, Kansas and Omaha, Nebraska) to send a message to a randomly selected target on the East Coast of the United States. Subjects were given minimal information about the target – their name and profession – and were instructed to send the message *directly* to the person, if they knew them, and, if they didn't know them, to send the message to someone they knew whom they thought would be more likely to know the target. Subjects recorded their details on a form accompanying the message, so that the path taken by the message could be reconstructed. Milgram found that the median length of the chain connecting

---

[69] This is the same Stanley Milgram whose famous experiments on obedience (Milgram, 1963) showed the ease with which ordinary people could, under the right conditions, be led to deliver electric shocks to a screaming subject at the request of an authority figure. That experiment, published two years after the trial of Adolf Eichmann in 1961, led to greater understanding of how normal people can be led to commit atrocities, as well as the rapid introduction of Institutional Review Boards and ethics panels for the purpose of protecting the well-being of research subjects.

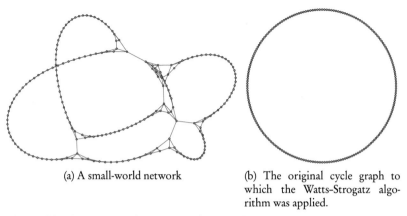

(a) A small-world network

(b) The original cycle graph to which the Watts-Strogatz algorithm was applied.

Figure 32 The average shortest path length of a small-world network differs considerably from that of the original graph to which the Watts–Strogatz algorithm was applied. The small-world network (a) has an average shortest path length of 12.45 whereas the cycle graph (b) has an average shortest path length of 25.38. Note how few edges need to be rewired in order to yield the 50 per cent reduction.

source to target was *five* with the modal value being *six*. This result has entered popular culture as the idea that "six degrees of separation" connect any two people on the planet.

The small-world network shown in Figure 32a was constructed using a process first described by Watts and Strogatz (1998), and later discussed in more detail by Watts (1999). The algorithm used to construct the graph is as follows. First, begin with a $1, 2$-circulant graph having N nodes. (The graph shown in Figure 31 is an example of such a graph; the embedding moves every other node inside a little bit so as to show the edge structure more clearly.) Second, each edge in the graph has a chance $p > 0$ of being 'rewired'. If an edge is selected for rewiring, the source of the edge is kept the same but the new target node is selected at random from the remaining nodes (excluding the source and original target). The graphs constructed by this algorithm are interesting because they have a property reminiscent of the 'small-world' phenomenon studied by Milgram: although the total number of edges is quite low compared to the number of possible edges, and the local structure of the network is quite regular, the network is nevertheless well connected in that the average shortest path length between any two randomly selected nodes is unusually low, given the sparsity of the graph.

Figure 33 shows the initial and final states for two different simulations of the Prisoner's Dilemma played on small-world networks. The payoffs used were the same ones considered earlier which were favourable to the spread of cooperation. However, here we see that the additional minor structure added

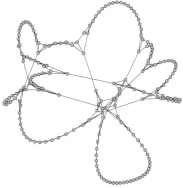

(a) A randomly chosen set of initial conditions for the Prisoner's Dilemma played on a small-world network of 200 nodes.

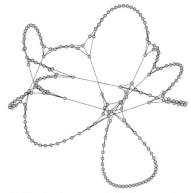

(b) The fixed state reached after a number of iterations, with all players using *imitate-the-best*.

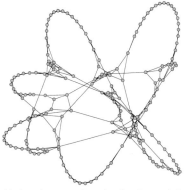

(c) Another set of randomly chosen initial conditions for the Prisoner's Dilemma played on a different small-world network of 200 nodes.

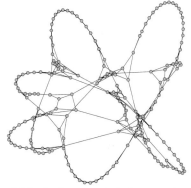

(d) The fixed state reached after a number of iterations, with all players using *imitate-the-best*.

◻ Cooperate ◼ Defect

Figure 33 Initial and final convergent states for the Prisoner's Dilemma played on two different small-world networks. The payoffs used were the same as in Figures 30 and 31.

by small-world networks *prevents* Cooperators from driving out most of the Defectors, contrary to what we saw happen in the case of 1, 2-circulant graphs, shown in Figure 31. Why does this happen?

We can see how the extra structure of small-world networks affects the evolutionary behaviour by using a different representation of the underlying graph. In Figure 34, the final states of the local interaction models from the previous figure are shown both as a force-layout embedding and as a circular embedding. What the circular embedding clearly shows is that the regions dominated by Cooperate are always bounded by edges which have been randomly rewired.

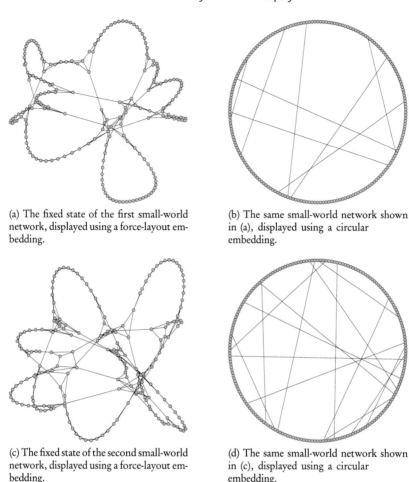

(a) The fixed state of the first small-world network, displayed using a force-layout embedding.

(b) The same small-world network shown in (a), displayed using a circular embedding.

(c) The fixed state of the second small-world network, displayed using a force-layout embedding.

(d) The same small-world network shown in (c), displayed using a circular embedding.

■ Cooperate   ■ Defect

Figure 34 Changing the representation of the underlying graph can reveal structure helpful for understanding the evolutionary behaviour.

This makes sense: local regions of the graph which have not been structurally affected by the random rewiring of an edge will have evolutionary dynamics exactly the same as an unaltered 1, 2-circulant graph: Cooperate will be capable of driving out Defect. But a randomly rewired edge introduces 'nonlocal' dependencies, in the sense that a Cooperator connected to a Defector via a rewired edge may find themselves interacting with a Defector whose local surroundings are more conducive to their persistence. Another way that rewiring can make a difference depends on whether a player was considered to be the *source* or *target* of a rewired edge. Players considered as the source of an edge will, after rewiring, still have the same number of interactions as before:

on a 1, 2-circulant graph, namely four. However, a player who was the target of a rewired edge will have that connection removed, reducing their number of interactions from four to three (and potentially fewer, if they were unlucky to be the target on multiple edges which were randomly selected for rewiring). And players who were selected as the new target for rewired edges will have one interaction more than most players in the population, and this can yield pay-offs for Defectors which prevent the spread of Cooperate. The important lesson here is that the possible evolutionary outcomes for small-world networks can be understood by studying evolutionary behaviour in the local regions unaffected by rewired edges, and then how the addition of rewired edges encourages or inhibits the spread of strategies.

Turning attention to two-dimensional lattice models, we find that spatial structure proves even more conducive to the survival and spread of cooperation than small-world networks. Figure 35 illustrates the ability of Cooperate to drive out Defection on a lattice where each player interacts with its eight nearest neighbours. The random starting conditions, featuring an unequal mix of Defect and Cooperate, initially seem to favour the strategy Defect. Isolated Defectors paired with eight Cooperating neighbours receive the highest possible payoff of the game, causing their strategy to be adopted by all of their neighbours. But this initial success forms sizeable clusters of Defect, where the subpar performance of Defect•—•Defect interactions gives regions of Cooperators an advantage when facing regions of Defectors.

The spread of Cooperate does not manage to drive out Defect entirely. As the irregular borders of advancing Cooperate regions come together, occasionally a Defector is well placed to take advantage of the surrounding change. By the fourth iteration, we see how stable 'islands' of Defect can persist as long as they are not too big. Isolated Defectors surrounded by Cooperators give rise to 'blinkers': the isolated Defector earns the highest possible payoof of the game, so in the next generation the isolated Defector expands to a 3×3 square. But that structure is unstable because all the periphery Defectors earn payoffs inferior to those of one of their Cooperate neighbours, and so the square collapses back to an isolated Defector.

Notice how radically different the situation is here compared with the sit-uation under the replicator dynamics. There, *any* amount of mutation would suffice to set a population along a path which would irreversibly lead to the All-Defect state. Here, the fact that Cooperate can largely drive out Defect in the absence of mutation shows that – at least for local interactions on the lattice – Cooperate can *persist* in the presence of small amounts of mutation, as Fig-ure 36 shows. The reason why is that if the mutation rate is sufficiently small, only a single Defector will be introduced into a region of Cooperators during

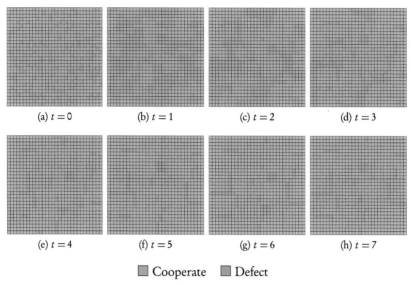

(a) $t = 0$ (b) $t = 1$ (c) $t = 2$ (d) $t = 3$

(e) $t = 4$ (f) $t = 5$ (g) $t = 6$ (h) $t = 7$

☐ Cooperate ☐ Defect

Figure 35 The spread of cooperation in the Prisoner's Dilemma played on a two-dimensional lattice. Interaction is with the eight nearest neighbours of the player (except at the edges, which are not assumed to wrap). The payoffs are the same as those which have previously been shown to be advantageous for the spread of cooperation.

any single generation. Although this may result in a shift of the local 'balance of power' between a few Defectors and Cooperators, the overall behaviour is not going to be overturned: regions of Cooperate soon achieve a steady state (except, perhaps, for some blinkers), perhaps with slightly adjusted borders.

4.2.3 The Centipede Game

Another model of cooperation which has frequently been studied is the *centipede game*. First introduced by Rosenthal (1981), the centipede game is in many respects a more interesting model of the problem of cooperation than the Prisoner's Dilemma because it involves multiple interactions between two players with aspects of *trust* included as well. Figure 37 shows the extensive-form representation of a six-stage centipede game. Play begins at the root node of the game tree (coloured white) with Player 1 having the first move. The strategies available at each node of the game tree are either to *Pass* control to the other player (thereby moving to the next node to the right in the game tree), or to *Take* the payoffs which are currently available (moving down, to a terminal node). The highest possible collective payoff occurs when both players choose Pass at each choice node, resulting in payoffs of 8 to Player 1 and 6 to Player 2.

But there is a curious feature built into the payoff structure of the centipede game, when analysed from the point of view of traditional game theory.

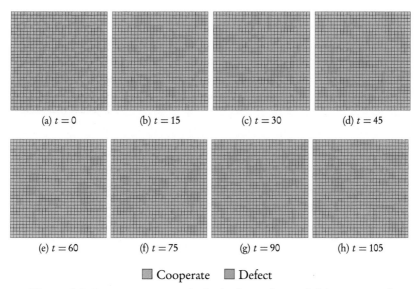

(a) $t = 0$ (b) $t = 15$ (c) $t = 30$ (d) $t = 45$

(e) $t = 60$ (f) $t = 75$ (g) $t = 90$ (h) $t = 105$

☐ Cooperate   ☐ Defect

Figure 36 Cooperate can persist in the face of non-trivial amounts of mutation. This figure shows snapshots taken every 15 iterations from the Prisoner's Dilemma played on a 30×30 lattice, Moore 8 neighbourhood, with a mutation rate of 5 per cent.

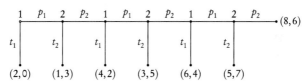

Figure 37 A six-stage extensive-form centipede game (from Smead, 2008).

Player 2, when inspecting the game tree, will notice that choosing Pass at the last node will yield a payoff of 6, whereas choosing Take will yield a payoff of 7. As a rational agent, player 2 thereby maximises his payoff by choosing Take at the last node. Since Player 1 is also a rational agent, she will have worked this out for herself. At her last decision node, she then knows that if she chooses Pass she will get a payoff of 5. Choosing Take, though, will yield a payoff of 6 instead. Player 1, then, will prefer to choose Take at her last decision node rather than Pass control to Player 2. Reasoning in this fashion leads to the counterintuitive result that rational agents, faced with the centipede game, will choose Take at the first node, yielding a payoff of 2 to Player 1 and a payoff of 0 to Player 2. And this happens *despite* the fact that the collectively optimal outcome, waiting at the end of the game tree, gives both players much more. The counterintuitive strategy of choosing Take immediately is the unique Nash equilibrium of the centipede game.

The centipede game gets its name because the structure used in Figure 37 can obviously be extended as far as you like. If the game had 100 choice nodes, thereby warranting its name, the final choice node for Player 2 is sufficiently far away from the start of the game that the perception triggering the reasoning process that leads to the counterintuitive outcome might not happen immediately. If Player 1 chooses Pass at the first move, they are taking a gamble that Player 2 will not immediately choose Take. In other words, Player 1 is *trusting* that Player 2 will behave pro-socially. But once play has moved onto the second choice node for Player 1, we are at a point where the payoffs to both players, regardless of what happens in the future, are strictly better than what traditional game theory predicts should happen.

What happens in local interaction models of the centipede game? We first need to specify two things: how we will handle the fact that the centipede game, unlike other games discussed so far in this section, is an *asymmetric* extensive form game, and how we will model a *strategy* for the centipede game. Let us consider each of these in turn.

The centipede game requires, for each pair (i,j) of individuals who play the game, that one be assigned the role of Player 1 and the other be assigned the role of Player 2. Perhaps the simplest way to assign roles is to do it by chance: given a pair (i,j), each person has a 50 per cent chance of being Player 1 and a 50 per cent of being Player 2. In the context of a local interaction model, this will be implemented as follows: for each round of play, each pair of individuals connected by an edge in the graph will be randomly assigned the roles of Player 1 and 2. The assignment of roles, for a given edge, will be independent of all other role assignments. This means that, for a given round of interactions, the *same individual* might find themselves in the role of Player 1 for some interactions, and in the role of Player 2 for other interactions. Finally, the assignment of roles on an edge will only apply for that *particular* round of play. In the next iteration, roles will be randomly reassigned.

Now that we have specified how roles will be assigned to players, how will *strategies* be modelled? As discussed in Section 1.3, extensive form games have, in principle, a huge number of potential strategies. This arises from the fact that a strategy in an extensive-form game specifies a course of action at each available choice node for a player, regardless of whether or not they are reached. But, as we saw for the game of tic-tac-toe, many of these strategies may lack internal consistency. Given this – especially since we are modelling players as boundedly rational actors – we can simplify the set of possible strategies by adopting a *behavioural* approach: a strategy simply specifies the stage in a centipede game where a player first chooses Take, on the assumption that the player will choose Pass at all earlier stages and will choose Take at all later

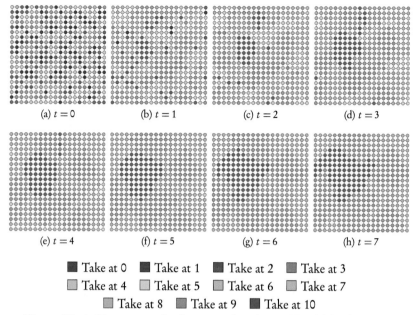

(a) $t = 0$ (b) $t = 1$ (c) $t = 2$ (d) $t = 3$

(e) $t = 4$ (f) $t = 5$ (g) $t = 6$ (h) $t = 7$

■ Take at 0 ■ Take at 1 ■ Take at 2 ■ Take at 3
☐ Take at 4 ☐ Take at 5 ☐ Take at 6 ☐ Take at 7
☐ Take at 8 ■ Take at 9 ■ Take at 10

Figure 38 A 10-stage centipede game played on a 20×20 lattice. Each player interacts with their four nearest neighbours at the cardinal compass points, with no wrapping at the edges and *imitate-the-best*.

stages, if they should happen to find themselves at that point of the game tree. Why is the latter qualification needed? It is needed because of how we assign roles! Suppose that an individual's strategy is to choose Take at stage 1 (i.e., adopting the traditional game-theoretic solution to the centipede game). If that individual is assigned the role of Player 2, their first opportunity to make a move won't occur until stage 2. And this off-by-one error can happen for any strategy. Suppose your strategy is to first choose Take at some stage k, where k is an even integer. If you are assigned the role of Player 1, all of your choice nodes occur at odd stages of the game. And the same thing happens if k is odd and you happen to be assigned the role of Player 2. Hence, a strategy specifies the stage at which an individual flips from being a trusting cooperator to someone who has decided that the game has gone on long enough and that they'd now better 'take the money and run.'

On a two-dimensional lattice, we find a stark demonstration of just how much of a difference the structure of local interactions makes. Figure 38 shows the outcome of the first seven generations of a 10-stage centipede game played on a lattice, beginning with random initial conditions. Over time, the strategies which spread under *imitate-the-best* are those which choose Pass in early stages of the game, leading to a greater collective payoff than the traditional game theoretic solution. The darkest colour present (in growing numbers) in iterations 6

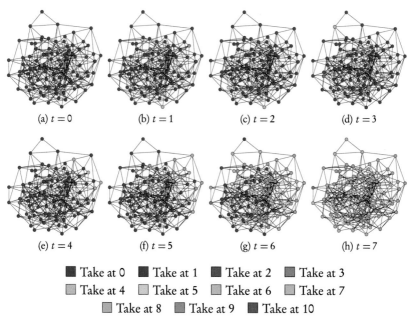

(a) $t = 0$ (b) $t = 1$ (c) $t = 2$ (d) $t = 3$

(e) $t = 4$ (f) $t = 5$ (g) $t = 6$ (h) $t = 7$

■ Take at 0   ■ Take at 1   ■ Take at 2   ■ Take at 3
□ Take at 4   □ Take at 5   □ Take at 6   □ Take at 7
□ Take at 8   ■ Take at 9   ■ Take at 10

Figure 39 The first seven iterations of the centipede game played on a random network of 100 individuals.

and 7 represents the strategy which goes to the rightmost point of the centipede game, choosing Pass at the very end *even though* it would be to Player 2's advantage to choose Take at that point. Why does this happen? There are two reasons. First, a player's strategy is used for *all* of their interactions in a given round of play, so they cannot condition their action on whether they are in the role of Player 1 or Player 2. Second, the assignment of roles is done randomly, so every player will be in the role of Player 1 or Player 2 approximately half of the time, each. Given this, it is beneficial for a *region* of players to all use the strategy that goes to the rightmost end of the centipede: although I *don't* benefit from doing so, when I am in the role of Player 2, I *do* benefit from you're doing so, when I am in the role of Player 1!

The ability of pro-social strategies to survive and persist in the centipede game does not depend on the regular structure of the lattice. Figure 39 illustrates the first seven iterations of the centipede game played on a random network of 100 individuals. The population was initialised to a state where everyone followed the Nash equilibrium strategy of choosing Take at the first stage of the game. With a mutation rate of 10 per cent,[70] we see that the population

---

[70] This is quite a high rate of mutation. However, since what generally matters for local interaction models is whether a local *cluster* of mutants happens to appear, this can also be thought of as speeding up time. With a low rate of mutation, one would have to wait a long time before

leaves the state where everyone follows the Nash equilibrium strategy to a state where the majority of the population choose Pass for many rounds of play. Some even follow the strategy which chooses Pass at every stage of the game. On random networks with some mutation, the population does not arrive at a fixed state. What happens, instead, is that the population moves between states where individuals tend to follow all of one strategy, sometimes going all the way to the end of the centipede game, sometimes not. But what *is* true is that the Nash equilibrium strategy of choosing Take immediately does not feature significantly beyond noise introduced by mutation.

Figure 40 shows aggregate results for the centipede game played on random networks (see Alexander, 2009). Each network contained 150 individuals,[71] and was randomly wired with each possible edge included with 3 per cent probability. The only constraint on the nature of the network was that it be *connected*, meaning that it was always the case that a path existed linking any two individuals together (perhaps via a lengthy sequence of intermediaries). Beginning with random initial conditions, each simulation was run for 300 generations (with no mutation). In Figure 40 the height of each bar represents the total number of individuals who employed each strategy after 300 generations. There are several things which are noteworthy. The first is that virtually no one chooses Take before stage 4. The second is that the distribution very roughly approximates results from the experimental literature. McKelvey and Palfrey (1992) found that approximately 15 per cent of subjects would choose the socially optimal outcome in the centipede game. In simulations, the socially optimal result occurs 13.3 per cent of the time.

Much more could be said about local interaction models and how the inclusion of social structure affects evolutionary games. Those interested in learning more, with a particular eye to philosophical applications, should consult Skyrms (2003) and Alexander (2007). The effect of finite populations for evolutionary games is covered in detail in Nowak (2006), as are many other topics which we were not able to consider. Many more games have been

---

a cluster of the right type appeared, which would then trigger a shift in the population state. With a higher rate of mutation, the waiting time is less because one eliminates many of the intervening generations where the population remains in stasis.

[71] Why 150 individuals? Dunbar (1992, 1993) argued that constraints on the cognitive capacities of humans place a natural ceiling on the number of meaningful social contacts we can have of around 150 persons. This value, known as 'Dunbar's number', has become a mainstay of popular science and folklore. However, recent research (see Lindenfors et al., 2021) suggests that the statistical methods used to derive this upper limit have such large confidence intervals that no meaningful value can be obtained.

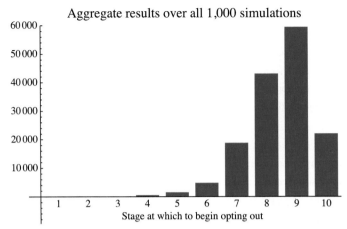

Figure 40 Aggregate results for the centipede game played on random networks.

studied using the local interaction framework than have been covered here, such as the stag hunt (a model of trust), the snowdrift game (another model of cooperation), the ultimatum game (a model of retribution), divide-the-dollar (a model of fair division), and signalling games (a model of the emergence of language). As one explores questions related to issues raised in this section – such as how boundedly rational agents learn from their experience, how dynamical models of network structure affect evolutionary outcomes, and how the very structure of a game, itself, can change over time – the boundaries between evolutionary game theory, decision theory, and epistemology become blurred in fruitful and fascinating ways.

If one had to summarise this section in a single sentence, I can think of no better statement than this: *the details of the evolutionary dynamics matter*. In particular, we have seen how finite population models can yield outcomes deeply at odds with the traditional game theoretical analyses. Complex regional cycling behaviour emerges in Rock-Paper-Scissors; strictly dominated strategies flourish in the Prisoner's Dilemma; and the unique Nash equilibrium identified by backwards induction in the centipede game is of no evolutionary significance. But it is important to keep in mind that all these results are contingent upon individuals using the learning rule of *imitate-the-best*. Changing the learning rule, or, for that matter, other parameters of the model, can make a enormous difference. Details of the evolutionary dynamics matter. The full answer of just how, and in just what ways, is the subject of ongoing research.

5 Conclusion

The White Rabbit put on his spectacles.
'Where shall I begin, please your Majesty?' he asked.
'Begin at the beginning,' the King said gravely,
'and go on till you come to the end: then stop.'
Lewis Carroll, Alice's Adventures in Wonderland

This Element has provided an introduction to some of the central topics in evolutionary game theory. It began with a discussion of why the traditional solution concept of game theory – the Nash equilibrium – is inadequate for capturing the concept of evolutionary stability. It then looked at the first, and most well-known, attempt to augment the Nash equilibrium solution concept with additional criteria: the *evolutionarily stable strategy* (ESS) solution concept by Maynard Smith and Price. We then considered several alternative characterisations of evolutionary stability, noting they all were logically equivalent to an ESS. This gave us reason to believe that the definition given for an ESS was not arbitrary, but rather captured core features of the idea of evolutionary stability.

It turned out, though, that while an ESS is a reasonably good first attempt at capturing the idea of evolutionary stability, it didn't cover all cases. In one example, we showed how it was possible for a *set* of strategies to behave in ways which seemed to warrant extending the concept of evolutionary stability to sets of strategies rather than just a single strategy. This led to the idea of *evolutionarily stable sets*.

At this point, things became complicated. First, because it turned out that there were several different ways to characterise the strength of the evolutionary stability of a set S. Should it be the case that *every* strategy in S can drive to extinction every strategy *not* in S? Or was it enough for every strategy not in S to be driven to extinction by at least one strategy in S? Second, there is then the further question of how *robust* the set S needs to be when it comes to driving out alternatives. For example, even if was the case that every strategy in S could drive to extinction every strategy not in S *provided that no more than one strategy not in S appeared at a time*, that might not seem a very important sense of evolutionary stability if the population was enormous. Suppose the Chinese government was given proof that their political system was assured to be stable provided that no more than one political dissident appeared at a time. Should they take comfort in this fact? This led us to turn away from *static* concepts of evolutionary stability to *dynamical* models of evolution.

A close study of dynamical models of evolution led to further surprising revelations. Whereas no weakly dominated strategies qualified as an ESS, under the replicator dynamics weakly dominated strategies could survive in great numbers. Moreover, the replicator dynamics allowed for asymptotically stable states to exist which didn't correspond to an ESS at all. And, if we considered population states which were not locally stable, but the inevitable state to which the population would converge, in the limit, we could get final outcomes which didn't even correspond to a Nash equilibrium of the underlying game.

Turning attention to other kinds of dynamical models – like the Brown–Nash–von Neumann (BNN) dynamics, or the Smith dynamics – led to other curious findings. Whereas both of these dynamical models were generated, at the population level, from individual belief revision processes which were arguably rational (to some degree), they allowed for the preservation of *strictly dominated* strategies.

As all three of these dynamical models – the replicator dynamics, the BNN dynamics, and the Smith dynamics – were continuous models in which arbitrarily small changes were meaningful and no population structure existed, in Section 4 we looked at finite population models. Here we found that the preservation of strictly dominated strategies was not at all uncommon: there was always some probability of it happening in the birth-death process, and it happened with great frequency in local interaction models of the Prisoner's Dilemma. But perhaps the most surprising feature uncovered was that, in local interaction models using *imitate-the-best*, we found pro-social behaviour surviving, in both the Prisoner's Dilemma and the centipede game, deviating significantly from the traditional game theoretic analyses of both games.

With such a wide variety of evolutionary models available where the predicted evolutionary outcome is so model-dependent, how is one supposed to know which models to use when? Although there are no absolute rules, a few general guidelines can be stated.

If all you have is the payoff matrix stating the strategic problem, and nothing else, then one of the static ESS concepts discussed in Section 2 is the place to begin. That is because all dynamical models make important additional assumptions about the kinds of interactions between individuals in the population and how individuals modify their strategy (or how reproduction takes place). But keep in mind that the ESS concept can identify a strategy which is unlikely to be explanatorily significant. Nowak (1990) provides an example of a game in which the ESS turns out to be dynamically inaccessible; as he said, 'evolutionarily stability does not imply that such a strategy will tend to evolve.'

If it is reasonable to assume that the population is very large and randomly mixing, then one of the continuous models from Section 3 may be the place to begin. (There are many other models, too!) Here, keep in mind how those models differ in the assumptions they make about the belief revision process used by individuals. Those assumptions about the epistemic state of individuals may or may not be warranted, depending on what information they have.

If it is reasonable to assume that the population is small and structured, then perhaps the approach of local interactions, from Section 4, is what you need. But keep in mind that local interaction models make a lot more assumptions than continuous models. In particular, they can be sensitive to the belief revision process, whether belief revision in the population occurs all at once or is staggered,[72] the way in which mutations occur, the network topology, and also whether the network topology is static or dynamic.[73] Those are a lot of assumptions which need to be justified! A good rule of thumb is to start simple and add complexity later, keeping in mind the quote from the British statistician George E. P. Box: 'All models are false, but some are useful.'

Evolutionary game theory is a fascinating field, full of mathematical subtleties and surprising results. Recent work has examined the emergence of language (see Skyrms, 2010), the emergence of unfairness (see O'Connor, 2019), and much more (see Sandholm, 2010). In the future, I suspect one main challenge will be connecting the results from the panoply of evolutionary game theoretic models to real-world *social* behaviour. Evolutionary game theory is a great source of *how-possible* explanations of social phenomena. The next step lies in converting some of those *how-possible* explanations into *how-actually* ones.

This is, perhaps, a fitting place to draw matters to a close. Evolutionary game theory came into existence through the realisation that game theory – the study of perfectly rational actors facing interdependent decision problems – could be applied to problems in evolutionary biology, despite the agents involved being very far removed from *rational* agents. The development of evolutionary game theory, over time, has led to a rich array of models and tools for studying boundedly rational actors in dynamic interdependent decision problems. Those types of actors, and those types of decision problems, are exactly the kind which underwrite our social existence. From the theory of games, to evolutionary biology, and back again, the intellectual trajectory of evolutionary game theory as a discipline has travelled full circle, returning to the social science origins from whence it began.

---

[72] See Huberman and Glance (1993) for a discussion of how this makes a difference in the spatial prisoner's dilemma.

[73] We didn't explore dynamic networks, but Skyrms and Pemantle (2000) and Spiekermann (2009) provide two different ways of modelling them.

References

Alexander, J. McKenzie (2007). *The Structural Evolution of Morality*. Cambridge University Press.

Alexander, J. McKenzie (2009). 'Social Deliberation: Nash, Bayes, and the Partial Vindication of Gabriele Tarde'. *Episteme* 6: 164–84.

Alexander, J. McKenzie (2016). 'Evolutionary Game Theory'. In William H. Batchelder, Hans Colonius, Ehtibar N. Dzhafarov and Jay Myung, eds., *New Handbook of Mathematical Psychology*, Cambridge University Press, volume 1, chapter 6, pp. 322–73.

Arneodo, Alain, Pierre Coulett, and Charles Tresser (1980). 'Occurence of Strange Attractors in Three-Dimensional Volterra Equations'. *Physics Letters* 79A: 259–63.

Artinger, Florian M., Gerd Gigerenzer and Perke Jacobs (2022). 'Satisficing: Integrating Two Traditions'. *Journal of Economic Literature* 60: 598–635.

Balkenborg, Dieter and Karl H. Schlag (2001). 'Evolutionarily Stable Sets'. *International Journal of Game Theory* 29: 571–95.

Barabási, Albert-László and Réka Albert (1999). 'Emergence of Scaling in Random Networks'. *Science* 286: 509–12.

Björnerstedt, Jonas and Jörgen Weibull (1999). 'Nash Equilibrium and Evolution by Imitation'. In Kenneth J. Arrow, Enrico Colombatto and Mark Perlman, eds., *The Rational Foundations of Economic Behavior*, St. Martin's Press.

Bomze, Immanuel M. (1983). 'Lotka-Volterra Equation and Replicator Dynamics: A Two-Dimensional Classification'. *Biological Cybernetics* 48: 201–11.

Brown, George W. and John von Neumann (1950). 'Solutions of Games by Differential Equations'. In *Contributions to the Theory of Games*, Princeton University Press, number 24 in Annals of Mathematical Studies.

Chao, Lin and Bruce R. Levin (1981). 'Structured Habitats and the Evolution of Anticompetitor Toxins in Bacteria'. *Proceedings of the National Academy of Sciences* 78: 6324–8.

Devaney, Robert L. (1989). *An Introduction to Chaotic Dynamical Systems*. Addison-Wesley Publishing Company, second edition.

Dunbar, Robin I. M. (1993). 'Coevolution of Neocortical Size, Group Size and Language in Humans'. *Behavioral and Brain Sciences* 16: 681–94.

Dunbar, Robin I. M. (1992). 'Neocortex Size as a Constraint on Group Size in Primates'. *Journal of Human Evolution* 22: 469–93.

Flood, Merrill M. (1952). 'Some Experimental Games'. Technical report, The RAND Corporation. RM-789-1.

Franchetti, Francisco and William H. Sandholm (2013). 'An Introduction to *Dynamo: Diagrams for Evolutionary Game Dynamics'. Biological Theory* 8: 167–78.

Gigerenzer, Gerd, Peter M. Todd and the ABC Research Group (1999). *Simple Heuristics That Make Us Smart*. Oxford University Press.

Gintis, Herbert (2009). *Game Theory Evolving*. Princeton University Press, second edition.

Gould, Stephen Jay (1980). *The Panda's Thumb: More Reflections in Natural History*. W. W. Norton & Company.

Harsanyi, John C. (1967a). 'Games with Incomplete Information Played by Bayesian Players. Part I. The Basic Model'. *Management Science* 14: 159–82.

Harsanyi, John C. (1967b). 'Games with Incomplete Information Played by Bayesian Players. Part II. Bayesian Equilibrium Points'. *Management Science* 14: 320–34.

Harsanyi, John C. (1967c). 'Games with Incomplete Information Played by Bayesian Players. Part III. The Basic Probability Distribution of the Game'. *Management Science* 14: 486–502.

Hilborn, Robert C. (2000). *Chaos and Nonlinear Dynamics: An Introduction for Scientists and Engineers*. Oxford University Press, second edition.

Hofbauer, Josef and Karl Sigmund (2002). *Evolutionary Games and Population Dynamics*. Cambridge University Press.

Hofbauer, Josef and William H. Sandholm (2011). 'Survival of Dominated Strategies under Evolutionary Dynamics'. *Theoretical Economics* 6: 341–77.

Hofbauer, Josef, Peter Schuster and Karl Sigmund (1979). 'A Note on Evolutionary Stable Strategies and Game Dynamics'. *Journal of Theoretical Biology* 81: 609–12.

Huberman, Bernardo A. and Natalie S. Glance (1993). 'Evolutionary Games and Computer Simulations'. *Proceedings of the National Academy of Sciences* 90: 7716–18.

Lindenfors, Patrik, Andreas Wartel and Johan Lind (2021). '"Dunbar's Number" Deconstructed'. *Biology Letters* 17: 202110158.

Luce, R. Duncan and Howard Raiffa (1957). *Games and Decisions: Introduction and Critical Survey*. John Wiley and Sons, Inc.

Maynard Smith, John and George Price (1973). 'The Logic of Animal Conflict'. *Nature* 246: 15–18.

McKelvey, Richard D. and Thomas R. Palfrey (1992). 'An Experimental Study of the Centipede Game'. *Econometrica* 60: 803–36.

Milgram, Stanley (1963). 'Behavioral Study of Obedience'. *Journal of Abnormal and Social Psychology* 67: 371–8.

Milgram, Stanley (1967). 'The Small World Problem'. *Psychology Today* 2: 61–7.

Mohseni, Aydin (2015). The Limits of Equilibrium Concepts in Evolutionary Game Theory. Master's thesis, Carnegie Mellon University.

Moran, Patrick A. P. (1958). 'Random Processes in Genetics'. In *Mathematical Proceedings of the Cambridge Philosophical Society*. Cambridge University Press, volume 54, pp. 60–71.

Nash, John (1950a). Non-cooperative Games. PhD thesis, Princeton University.

Nash, John F. (1950b). 'Equilibrium Points in n-Person Games'. *Proceedings of the National Academy of Sciences* 36: 48–9.

Nowak, Martin A. (1990). 'An Evolutionary Stable Strategy May Be Inaccessible'. *Journal of Theoretical Biology* 142: 237–41.

Nowak, Martin A. (2006). *Evolutionary Dynamics: Exploring the Equations of Life*. Harvard University Press.

O'Connor, Cailin (2019). *The Origins of Unfairness: Social Categories and Cultural Evolution*. Oxford University Press.

Oechssler, Jörg (1997). 'An Evolutionary Interpretation of Mixed-Strategy Equilibria'. *Games and Economic Behavior* 21: 203–37.

Pacuit, Eric and Olivier Roy (2017). 'Epistemic Foundations of Game Theory'. In E. N. Zalta, ed., *The Stanford Encyclopedia of Philosophy*, Metaphysics Research Lab, Stanford University. Summer 2017 edition.

Rosenthal, Robert W. (1981). 'Games of Perfect Information, Predatory Pricing, and the Chain Store'. *Journal of Economic Theory* 25: 92–100.

Rubinstein, Ariel (1991). 'Comments on the Interpretation of Game Theory'. *Econometrica* 59: 909–24.

Rubinstein, Ariel (1998). *Modeling Bounded Rationality*. Massachusetts Institute of Technology Press.

Samuelson, Larry (1996). 'Bounded Rationality and Game Theory'. *The Quarterly Review of Economics and Finance* 36, Supplement 1: 17–35.

Sandholm, William H. (2010). *Population Games and Evolutionary Dynamics*. Massachusetts Institute of Technology Press.

Sandholm, William H. (2015). 'Population Games and Deterministic Evolutionary Dynamics'. In H. P. Young and Shmuel Zamir, eds., *Handbook of Game Theory*, Elsevier, volume 4, chapter 13, pp. 720–78.

Simon, Herbert A. (1955). 'A Behavioral Model of Rational Choice'. *The Quarterly Journal of Economics* 69: 99–118.

Simon, Herbert A. (1956). 'Rational Choice and the Structure of the Environment'. *Psychological Review* 63: 129.

Skyrms, Brian (1990). *The Dynamics of Rational Deliberation*. Harvard University Press.

Skyrms, Brian (1992). 'Chaos in Game Dynamics'. *Journal of Logic, Language, and Information* 1: 111–30.

Skyrms, Brian (1993). 'Chaos and the Explanatory Significance of Equilibrium: Strange Attractors in Evolutionary Game Dynamics'. In *Proceedings of the 1992 PSA*, volume 2, pp. 374–94.

Skyrms, Brian (2003). *The Stag Hunt and the Evolution of Social Structure*. Cambridge University Press.

Skyrms, Brian (2010). *Signals: Evolution, Learning, & Information*. Oxford University Press.

Skyrms, Brian and Robin Pemantle (2000). 'A Dynamic Model of Social Network Formation'. *Proceedings of the National Academy of Science* 97: 9340–6.

Smead, Rory (2008). 'The Evolution of Cooperation in the Centipede Game with Finite Populations'. *Philosophy of Science* 75: 157–77.

Smith, Michael J. (1984). 'The Stability of a Dynamic Model of Traffic Assignment: An Application of a Method of Lyapunov'. *Transportation Science* 18: 245–52.

Spiekermann, Kai P. (2009). 'Sort Out Your Neighbourhood: Public Good Games on Dynamic Networks'. *Synthese* 168: 273–94.

Taylor, Peter D. and Leo B. Jonker (1978). 'Evolutionary Stable Strategies and Game Dynamics'. *Mathematical Biosciences* 40: 145–56.

Thomas, Bernhard (1985). 'On Evolutionarily Stable Sets'. *Journal of Mathematical Biology* 22: 105–15.

Vanderschraaf, Peter (2006). 'War or Peace? A Dynamical Analysis of Anarchy'. *Economics & Philosophy* 22: 243–79.

Vanderschraaf, Peter (2007). 'Covenants and Reputations'. *Synthese* 157: 167–95.

Vickers, Glenn T. and Chris Cannings (1987). 'On the Definition of an Evolutionarily Stable Strategy'. *Journal of Theoretical Biology* 129: 349–53.

von Neumann, John (1928). 'Zur Theorie der Gesellschaftsspiele'. *Mathematische Annalen* 100: 295–320. Translated by S. Bargmann as 'On the Theory of Games of Strategy', in *Contributions to the Theory of Games*, volume 4, *Annals of Mathematics Studies*, 40, pp. 13–43, 1959.

Watts, Duncan J. (1999). *Small Worlds: The Dynamics of Networks between Order and Randomness*. Princeton University Press.

Watts, Duncan J. and Steven H. Strogatz (1998). 'Collective Dynamics of 'Smallworld' Networks'. *Nature* 393: 440–2.

Weibull, Jörgen W. (1995). *Evolutionary Game Theory*. Massachusetts Institute of Technology Press.

Zeeman, Erik C. (1979). 'Population Dynamics from Game Theory'. In *Global Theory of Dynamical Systems: Proceedings of an International Conference Held at Northwestern University, Evanston, Illinois, June 18–22, 1979*. Northwestern: University Press.

Acknowledgements

Many thanks to Martin Peterson, the editor of this Cambridge Elements series, for his near-infinite patience in seeing this work through to completion, as a delay of a couple months turned into a delay of a couple years. In this case, matters were compounded by the fact that I had originally planned to write this element during the 2019–20 academic year, while also serving as Head of Department. Needless to say, that plan didn't work out terribly well due to the world being turned upside down by a piece of zoonotically transmitted bat RNA.

I want to dedicate this element to my parents, for all of their love and support over the years. They are causally responsible for this element, in two different ways. The first is obvious. The second is due to that first computer they gave me in the early 1980s, which unexpectedly sparked the interest in the mathematics and programming found within these pages. Thank you both so much, for everything.

Cambridge Elements ≡

Decision Theory and Philosophy

Martin Peterson
Texas A&M University

Martin Peterson is Professor of Philosophy and Sue and Harry E. Bovay Professor of the History and Ethics of Professional Engineering at Texas A&M University. He is the author of four books and one edited collection, as well as many articles on decision theory, ethics and philosophy of science.

About the Series

This Cambridge Elements series offers an extensive overview of decision theory in its many and varied forms. Distinguished authors provide an up-to-date summary of the results of current research in their fields and give their own take on what they believe are the most significant debates influencing research, drawing original conclusions.

Cambridge Elements $\equiv$

Decision Theory and Philosophy

Printed in the United States
by Baker & Taylor Publisher Services